EATCS

Monographs on Theoretical Computer Science

Volume 8

EATCS Monographs on Theoretical Computer Science

Vol. 1: K. Mehlhorn: Data Structures and Algorithms 1: Sorting and Searching. XIV, 336 pages, 87 figs. 1984.

Vol. 2: K. Mehlhorn: Data Structures and Algorithms 2: Graph Algorithms and NP-Completeness. XII, 260 pages, 54 figs. 1984.

Vol. 3: K. Mehlhorn: Data Structures and Algorithms 3: Multi-dimensional Searching and Computational Geometry. XII, 284 pages, 134 figs. 1984.

Vol. 4: W. Reisig: Petri Nets. An Introduction. X, 161 pages, 111 figs. 1985.

Vol. 5: W. Kuich, A. Salomaa: Semirings, Automata, Languages. IX, 374 pages, 23 figs. 1986.

Vol. 6: H. Ehrig, B. Mahr: Fundamentals of Algebraic Specification 1. Equations and Initial Semantics. XI, 321 pages, 1985.

Vol. 7: F. Gécseg: Products of Automata. VIII, 107 pages, 18 figs. 1986.

Vol. 8: F. Kröger: Temporal Logic of Programs. VIII, 148 pages, 1987

Fred Kröger

Temporal Logic of Programs

Springer-Verlag Berlin Heidelberg New York
London Paris Tokyo

Editors
Prof. Dr. Wilfried Brauer
Institut für Informatik, Technische Universität München
Arcisstr. 21, D-8000 München 2, Germany

Prof. Dr. Grzegorz Rozenberg
Institute of Applied Mathematics and Computer Science
University of Leiden, Wassenaarseweg 80, P.O. Box 9512
NL-2300 RA Leiden, The Netherlands

Prof. Dr. Arto Salomaa
Department of Mathematics, University of Turku
SF-20500 Turku 50, Finland

Author
Prof. Dr. Fred Kröger
Institut für Informatik der Universität München
Theresienstr. 39, D-8000 München 2

ISBN-13:978-3-642-71551-8 e-ISBN-13:978-3-642-71549-5
DOI: 10.1007/978-3-642-71549-5

Library of Congress Cataloging in Publication Data.
Kröger Fred, 1945–
Temporal logic of programs.
(EATCS monographs on theoretical computer science; v. 8)
Bibliography: p.
Includes index.
1. Electronic digital computers – Programming.
2. Logic, Symbolic and mathematical. I. Title. II. Series.
QA76.6.K753 1987 005.13'1 86-31336
ISBN-13:978-3-642-71551-8

Softcover reprint of the hardcover 1st edition 1987

Typesetting: Universitätsdruckerei H. Stürtz AG, Würzburg
2145/3020-543210

Preface

Temporal logic is a logic of propositions whose truth and falsity may depend on time. Closely related to modal logics, it has been studied for a long time. Precise formal foundations of (various kinds of) temporal logic have been laid during the last, say, 25 years.

In classical mathematics propositions do not depend on time (they are *static* in some sense), so temporal logic is not of much interest there. The mathematical treatment of *programs*, however, contains a significant *dynamic* aspect. A typical model of the execution of a program is a sequence of *states* (in the "flow of time"). In different states, program entities such as variables may have different values and, hence, propositions about these values may have different truth values. *Temporal logic of programs* means taking temporal logic as a basis for the mathematics of execution sequences of programs and applying the logical means – language and deduction mechanisms – to the formal description and analysis of dynamic program properties.

From its very beginning about 10 years ago, this approach has received much attention and been a remarkable success, in particular in the field of parallel programs. It still constitutes a large area of present-day research. This monograph – an elaboration of the notes from courses given in the winter semesters 1983/84 and 1984/85 at the Technical University of Munich – tries to give a comprehensive and uniform presentation of some of the material which has been developed during the last few years and has now apparently reached some sort of "saturated state". Moreover, putting together the various notions, methods, results and applications we have paid much attention to precisely elaborated proofs and intuitive motivations and explanations of technical details. So the book may also serve as a textbook for graduate students and teachers.

The content of the book is divided into three main parts. Chapters I–III deal with the pure ("linear time") temporal logic. In Chapter I the new linguistic features with their formal semantics are introduced. In Chapter II the new (propositional) logic is axiomatized, and in Chapter III it is extended to a first-order logic.

Chapter IV describes the "temporal semantics" of (parallel) programs, i.e., how to represent programs and their properties within the language of temporal logic.

Chapters V–VII present applications of the logical apparatus to the verification of program properties. Chapter V deals with *invariance* and *precedence* properties; *liveness* properties are treated in Chapter VI. The development of the basic proof methods in Sections 16, 18, 21 and 22 is accompanied by various elaborated examples for their application in Sections 17, 19 and 23. A discussion of special methods for sequential programs in Chapter VII concludes this third part.

The reader of this book is assumed to be familiar with the general concepts of mathematical logic and the main concepts of classical propositional and first-order logic. In the Introduction we give only a short summary of those notions which will be needed in the following.

The book presents material from many sources. We have not included all the corresponding citations in the text, but in separate bibliographical remarks at the end of the book we have tried to give – hopefully – the proper credit to everyone whose publications have been used.

I am grateful to A. Kausche, K. Klaus, H. Schlingloff, F. Stolz and – in particular – H. Vogel who helped in debugging preliminary versions of this monograph and provided important suggestions for several improvements.

I am also indebted to A. Bussmann and U. Weber for their carefulness and patience when typing the manuscript and its various alterations and extensions.

Finally, I would like to thank the editors of this monograph series – in particular W. Brauer – and Springer-Verlag for their interest in my manuscript and the support during the completion of this volume.

Munich, November 1986 Fred Kröger

Contents

Introduction

Logic and Programs

Important goals of mathematical logic are, in general, to:

- provide languages for the precise formulation of *propositions*,
- investigate mechanisms for finding out the *truth* or *falsity* of propositions.

In the view of classical (propositional or first-order) logic, a proposition is a "sentence" for which it makes sense to ask whether it is true or false, for example,

"3 divides 8",
"Every man is mortal",

but not:

"Go to the door".

Consider now the following sentence:

"Today it's raining".

It makes sense to ask whether this sentence is true or not. However, there is a new aspect: the answers "true" or "false" may be different on different days. We may say that this is a proposition the truth values of which depend on *time*.

It is the goal of *temporal logic* to investigate languages and "logical instruments" for just such propositions and their temporal relationships.

What has temporal logic to do with computer programs? Consider the following fragment of such a program:

$$\begin{array}{ccccccc} \ldots\ ; & c:=b\ ; & b:=b-a & ; & \ldots \\ \uparrow & \uparrow & & \uparrow & \\ \alpha & \beta & & \gamma & \end{array}$$

α, β and γ point to places in the program and can also be viewed as *states* (time points!) which are passed through when executing this sequence of assignments. Now let A denote the proposition:

$$(a+b=c \wedge a>0) \rightarrow b>0$$

(in usual first-order notation) and assume that the variables a, b, c have the values 3, -3, 0, respectively, when execution is in state α. With these values A is false and thus:

A is false at α.

In state β, reached by executing $c := b$ at α, we have $a=3$, $b=-3$, $c=-3$ and this implies:

A is true at β.

At γ we then have $a=3$, $b=-6$, $c=-3$ and therefore:

A is false at γ.

This simple example shows that certain assertions about a program (e.g., relationships between the program variables) can be viewed as propositions in the temporal logic sense – depending on execution states the sequence of which plays the role of "time". Assertions of this kind can be used to describe interesting properties of programs (e.g., "correctness") and since they fit into the abstract model of temporal logic we can try to apply this logic to the description and investigation of such program properties.

Temporal logic of programs – in this sense – has been developed into a powerful tool and constitutes a large field of present-day research. This book gives a comprehensive presentation of this theory – or better: of that part of it which is now sometimes called the *Manna-Pnueli theory* of temporal program logic. There are many other interesting aspects within the broad general topic which are not included and are only referred to by giving some relevant literature. Mainly:

- We consider neither "branching time" temporal logic, nor "interval logic". We also do not follow up most recent investigations of "compositionality" of temporal program logic and of linguistic extensions by "past" operators.
- We consider only a special class of programs. The description of other "systems" is not dealt with.
- We do not deal with the field of temporal "specification" of dynamic systems.

In many parts we will present the material according to patterns given in the relevant literature, mainly the work of Manna and Pnueli (1982a, b, c, 1983b, c), but we also introduce new concepts, presentations and aspects, partly guided by personal taste but also aiming at a theory as elegant, applicable and precise as possible. Some catchwords of our representation are:

- We use a new basis of logical operators including the atnext operator recently introduced by the author.
- The linguistic elements for the description of programs are slightly extended by introducing formulas expressing that "an action is executed" besides the usual description of "an action is ready to execute".
- We investigate structured programs instead of unstructured "transition graphs".
- Particular attention is paid to a detailed elaboration of the "temporal semantics" of programs.
- Program verification principles are fully formalized within the logical language.
- It is demonstrated that these verification methods are based on respective purely logical proof principles and on some minimal information about the program semantics.
- It is shown how Hoare's partial correctness calculus can be embedded into the temporal framework.

Historical Remarks

Temporal logic, as we want to describe it, is a branch of *modal logic* which has been studied for a long time. Modal logic deals with two propositional operators $\Box$ and $\Diamond$ (in addition to the usual ones like $\wedge$, $\vee$, $\rightarrow$, etc.) interpreted as "necessarily" and "possibly". This is based on the idea that the truth of an assertion is a relative notion depending on *possible worlds*. A formal semantics was presented in this way by Kripke (1963). Prior (1957) was the first to suggest a "temporal" interpretation of $\Box$ and $\Diamond$: "always" and "sometime". In the sequel to this, many different systems of temporal logic were studied and an overview of these developments can be found in Rescher and Urquhart (1971). It should be noticed that in these contexts temporal logic as we want to do it is usually called *tense logic* whereas the term "temporal logic" is used differently.

Of special interest for us is temporal (or tense) logic assuming a discrete and linearly ordered time structure. A logic with $\Box$ and $\Diamond$ over this model is equivalent to the modal system S4.3.1. However, the concept of linearity also made possible the investigation of new temporal operators concerning the notion of *next time*. v. Wright (1965) suggested a logic based on a binary operator "and next"; in v. Wright (1966) he introduced the binary operator "and then". The combination of these systems in v. Wright (1967) led to a logic with the operators "always" and "next" (the *nexttime operator*). A complete formal system for the latter logic was first given by Prior (1967) who also suggests using such systems for proofs of the "working of digital computers". Prior attributes the formal system to Lemmon. Probably it should appear in Lemmon (1966) but Lemmon died before finishing this book. Other similar systems were given by Scott (reported in Prior (1967)), Clifford (1966), and Segerberg (1967). Finally, Kamp (1968) introduced the binary operator "until". Complete axiomatizations for this operator and its counterpart "since" can be found in Burgess (1982).

A first concrete mention of how the modal operators "always" and "sometime" could be used in program verification was given by Burstall (1974). This idea was elaborated by Pnueli (1977). The present author suggested a somewhat different approach in Kröger (1975) with an operator "and then" (somewhat different from v. Wright's "and next") modelling in some sense the concatenation of program statements and a complicated "loop" operator. This idea was elaborated in Kröger (1976 and 1977), where an operator "next" is also used. The combination of the operators "next", "always" and "sometime" in the field of verification of (sequential) programs was – to our knowledge – first investigated in Kröger (1978). Pnueli (1979) essentially improved the semantical apparatus of this logic, gave a finitary proof system for it (in contrast to the infinitary one in our paper) and extended its application to parallel programs. From that time on, a huge number of investigations arose and the development seems by no means yet finished. Some more remarks on the more recent literature are contained in a particular section at the end of the book.

Some Concepts and Notions of Classical Logic

A *logical language* is given by an alphabet of symbols and the definition of a set of strings over this alphabet, called *formulas*. The simplest kind of such a language is a language $\mathscr{L}_A$ of (*classical*) *propositional logic* (the index A denotes the German word "Aussagenlogik") which can be given as follows.

Alphabet

- A denumerable set $\mathscr{V}$ of *atomic formulas*,
- the symbols $\neg$, $\rightarrow$, (,).

Inductive definition of *formulas*

1. Every atomic formula is a formula.
2. If A is a formula then $\neg A$ is a formula.
3. If A and B are formulas then $(A \rightarrow B)$ is a formula.

An inductive definition may be understood to be like a set of production rules of a formal grammar: a string over the alphabet is a formula if and only if it can be "produced" by finitely many applications of the rules 1–3.

Further logical operators and constants can be introduced to abbreviate particular formulas.

Abbreviations

$A \wedge B$	for $\neg(A \rightarrow \neg B)$,
$A \vee B$	for $\neg A \rightarrow B$,
$A \leftrightarrow B$	for $(A \rightarrow B) \wedge (B \rightarrow A)$,
`true`	for $v_0 \vee \neg v_0$ (with some particular $v_0 \in \mathscr{V}$)
`false`	for $\neg$ `true`

(We have omitted surrounding parentheses.)

The symbols A and B in such formulations are not formulas themselves but *syntactic variables* ranging over the set of formulas.

The *semantics* of such a language $\mathscr{L}_A$ is based on the concept of (*Boolean*) *valuations*: a valuation $\mathbf{B}$ is a mapping

$$\mathbf{B}: \mathscr{V} \rightarrow \{\mathbf{f}, \mathbf{t}\}$$

where **f** and **t** are called *truth values* (representing "false" and "true", respectively). Every **B** can be inductively extended to the set of all formulas:

1. $\mathbf{B}(v)$ for $v \in \mathscr{V}$ is given.
2. $\mathbf{B}(\neg A) = \mathbf{t}$ iff $\mathbf{B}(A) = \mathbf{f}$.
3. $\mathbf{B}(A \rightarrow B) = \mathbf{t}$ iff $\mathbf{B}(A) = \mathbf{f}$ or $\mathbf{B}(B) = \mathbf{t}$.

This also defines **B** for the other operators, for example,

$$\mathbf{B}(A \wedge B) = \mathbf{t} \quad \text{iff } \mathbf{B}(A) = \mathbf{t} \quad \text{and} \quad \mathbf{B}(B) = \mathbf{t}.$$

A formula A is called *valid in* $\mathbf{B}$ (denoted by $\Vdash_{\mathbf{B}} A$) if $\mathbf{B}(A)=\mathbf{t}$. A is called *valid* or *tautology* (denoted by $\Vdash A$) if $\Vdash_{\mathbf{B}} A$ holds for every $\mathbf{B}$. A *follows from* a set $\mathscr{F}$ of formulas (denoted by $\mathscr{F} \Vdash A$) if $\Vdash_{\mathbf{B}} A$ holds for every $\mathbf{B}$ with $\Vdash_{\mathbf{B}} B$ for all $B \in \mathscr{F}$.

Obviously, there are two notions of "logical consequence" in $\mathscr{L}_{\mathrm{A}}$. The first one is expressed by the *implication* operator $\rightarrow$ within the language:

$$A \rightarrow B.$$

The second one is given by the relation $\Vdash$:

$$A \Vdash B.$$

(We write "A" instead of "$\{A\}$".) A fundamental fact of classical logic is that these notions are equivalent:

$$A \Vdash B \quad \text{iff} \quad \Vdash A \rightarrow B$$

or, more generally:

$$A_1, \ldots, A_n \Vdash B \quad \text{iff} \quad \Vdash (A_1 \wedge \ldots \wedge A_n) \rightarrow B.$$

The valid formulas of $\mathscr{L}_{\mathrm{A}}$ can also be characterized by a formal system. A *formal system* Σ for some logical language $\mathscr{L}$ consists of

- a set of formulas of $\mathscr{L}$, called *axioms*,
- *rules* of the form $A_1, \ldots, A_n \vdash B \quad (n \geq 1)$.

The formulas $A_1, \ldots, A_n$ are called the *premises*, the formula B is the *conclusion* of this rule. The *derivability* of a formula A in Σ (denoted by $\vdash_{\Sigma} A$ or simply $\vdash A$) is defined inductively:

1. Every axiom is derivable.
2. If the premises of a rule are derivable then the conclusion of this rule is derivable.

A formula A is called *derivable from* a set $\mathscr{F}$ of formulas (denoted by $\mathscr{F} \vdash_{\Sigma} A$ or $\mathscr{F} \vdash A$) if A is derivable in the formal system which results from Σ by taking all formulas of $\mathscr{F}$ as additional axioms. This implies that:

$$\vdash A \quad \text{iff} \quad \emptyset \vdash A.$$

If A is derivable from some $A_1, \ldots, A_n$ then the "relation" $A_1, \ldots, A_n \vdash A$ can itself be used as a *derived rule* in other derivations.

There are many possible formal systems for $\mathscr{L}_{\mathrm{A}}$. We note the following system Σ_{A}:

Axioms

- $A \rightarrow (B \rightarrow A)$,
- $(A \rightarrow (B \rightarrow C)) \rightarrow ((A \rightarrow B) \rightarrow (A \rightarrow C))$,
- $(\neg A \rightarrow \neg B) \rightarrow (B \rightarrow A)$.

Rule

- $A, A \rightarrow B \vdash B$ (*modus ponens*).

We remark once more that the strings written down are not formulas. So, e.g., $A \to (B \to A)$ is not really one axiom but an axiom scheme which yields infinitely many axioms when substituting formulas for A and B.

Like the semantic relation $\Vdash$, derivability is "equivalent" to implication in the following sense:

$$A \vdash B \quad \text{iff} \quad \vdash A \to B.$$

The only if part of this fact is the so-called *deduction theorem*.

Actually the relations $\vdash_{\Sigma_A}$ and $\Vdash$ themselves are equivalent:

$$\mathscr{F} \vdash_{\Sigma_A} A \quad \text{iff} \quad \mathscr{F} \Vdash A$$

which implies also:

$$\vdash_{\Sigma_A} A \quad \text{iff} \quad \Vdash A.$$

This fact is called *soundness* and *completeness* of Σ_A.

Propositional logic investigates the logical operations $\neg$, $\to$, $\vee$, $\wedge$, etc., starting from some atomic formulas of which no further details are given. *First-order predicate logic* is based on propositional logic but additionally looks closer at the structure of atomic formulas and allows *quantification.*

A (*classical*) *first-order language* $\mathscr{L}_P$ is given as follows.

Alphabet

- Denumerably many (*subject*) *variables*,
- for every $n \in \mathbb{N}_0$, at most denumerably many *n-ary function symbols* (also called *constants* in the case $n=0$),
- for every $n \in \mathbb{N}_0$, $n \geq 1$, at most denumerably many *n-ary predicate symbols*,
- the binary predicate symbol $=$,
- the symbols $\neg$, $\to$, $\forall$, (,).

($\mathbb{N}_0$ denotes the set of natural numbers including 0.)

Inductive definition of *terms*

1. Every variable is a term.
2. If f is an n-ary function symbol and $t_1, \ldots, t_n$ are terms then $f(t_1, \ldots, t_n)$ is a term.

An *atomic formula* is a string of the form $p(t_1, \ldots, t_n)$ where p is an n-ary predicate symbol and $t_1, \ldots, t_n$ are terms. We write $t_1 = t_2$ instead of $=(t_1, t_2)$.

Inductive definition of *formulas*

1. Every atomic formula is a formula.
2. If A and B are formulas then $\neg A$ and $(A \to B)$ are formulas. (We write $t_1 \neq t_2$ instead of $\neg t_1 = t_2$.)
3. If A is a formula and x is a variable then $\forall x A$ is a formula.

In addition to the abbreviations as in $\mathscr{L}_A$ we may introduce:

$\exists x A$ for $\neg\forall x \neg A$.

The occurrence of a variable x in some formula A is called *bound* if it appears in some part $\forall x B$ of A. Otherwise it is called *free*. If t is a term then $A_x(t)$ denotes the result of substituting t for every free occurrence of x in A. When writing $A_x(t)$ we always assume implicitly that t does not contain a variable which occurs bound in A. (This can always be achieved by replacing the bound variables of A by others.)

The basic semantical concept of first-order logic is the following:
A *structure* **S** for $\mathscr{L}_P$ consists of

- a set $|\mathbf{S}| \neq \emptyset$, called *universe*,
- an n-ary function $\mathbf{S}(f): |\mathbf{S}|^n \to |\mathbf{S}|$ for every n-ary function symbol f,
- an n-ary relation $\mathbf{S}(p) \subset |\mathbf{S}|^n$ for every n-ary predicate symbol p other than $=$.

A *variable valuation* ξ (with respect to **S**) assigns some $\xi(x) \in |\mathbf{S}|$ to every variable x of $\mathscr{L}_P$. A structure together with a variable valuation ξ defines a value $\mathbf{S}^{(\xi)}(t) \in |\mathbf{S}|$ for every term t:

1. $\mathbf{S}^{(\xi)}(x) = \xi(x)$ for every variable x.
2. $\mathbf{S}^{(\xi)}(f(t_1, \ldots, t_n)) = \mathbf{S}(f)(\mathbf{S}^{(\xi)}(t_1), \ldots, \mathbf{S}^{(\xi)}(t_n))$.

Furthermore, we can define $\mathbf{S}^{(\xi)}(A) \in \{\mathbf{f}, \mathbf{t}\}$ for every atomic formula:

1. $\mathbf{S}^{(\xi)}(p(t_1, \ldots, t_n)) = \mathbf{t}$ iff $(\mathbf{S}^{(\xi)}(t_1), \ldots, \mathbf{S}^{(\xi)}(t_n)) \in \mathbf{S}(p)$ for every p other than $=$.
2. $\mathbf{S}^{(\xi)}(t_1 = t_2) = \mathbf{t}$ iff $\mathbf{S}^{(\xi)}(t_1) \underset{|\mathbf{S}|}{=} \mathbf{S}^{(\xi)}(t_2)$, where $\underset{|\mathbf{S}|}{=}$ denotes equality in $|\mathbf{S}|$.

Now $\mathbf{S}^{(\xi)}$ plays the role of the valuation **B** in $\mathscr{L}_A$ and can be inductively extended to all formulas:

1. $\mathbf{S}^{(\xi)}(A)$ for atomic formulas is already defined.
2. $\mathbf{S}^{(\xi)}(\neg A) = \mathbf{t}$ iff $\mathbf{S}^{(\xi)}(A) = \mathbf{f}$.
3. $\mathbf{S}^{(\xi)}(A \to B) = \mathbf{t}$ iff $\mathbf{S}^{(\xi)}(A) = \mathbf{f}$ or $\mathbf{S}^{(\xi)}(B) = \mathbf{t}$.
4. $\mathbf{S}^{(\xi)}(\forall x A) = \mathbf{t}$ iff $\mathbf{S}^{(\xi')}(A) = \mathbf{t}$ for every ξ' with $\xi'(y) \underset{|\mathbf{S}|}{=} \xi(y)$ for every y other than x.

A formula A of $\mathscr{L}_P$ is called *valid in* **S** ($\Vdash_{\mathbf{S}} A$) if $\mathbf{S}^{(\xi)}(A) = \mathbf{t}$ for every ξ. A is called *valid* ($\Vdash A$) if $\Vdash_{\mathbf{S}} A$ for every **S**. A *follows from* a set $\mathscr{F}$ of formulas ($\mathscr{F} \Vdash A$) if $\Vdash_{\mathbf{S}} A$ for every **S** with $\Vdash_{\mathbf{S}} B$ for every $B \in \mathscr{F}$.

Again there exist sound and complete formal systems for $\mathscr{L}_P$. An example is the following system Σ_P:

Axioms

- All axioms of $\Sigma_{\mathbf{A}}$,
- $\forall x A \rightarrow A_x(t)$,
- $x = x$,
- $x = y \rightarrow (A \rightarrow A_x(y))$.

Rules

- $A, A \rightarrow B \vdash B$,
- $A \rightarrow B \vdash A \rightarrow \forall x B$ if there is no free occurrence of x in A (*generalization*).

Chapter I
Propositional Temporal Logic

1. A Language $\mathscr{L}_{\mathrm{TA}}$ of Propositional Temporal Logic

Let us begin with an informal discussion of the basic ideas already indicated in the introduction. The starting point of our considerations is the idea that there are different *time points* which may yield different truth values of propositions. We make a first important agreement: we assume the set of time points to be infinite, discrete and *linearly ordered* with a smallest element. This leads to the following picture of a time scale:

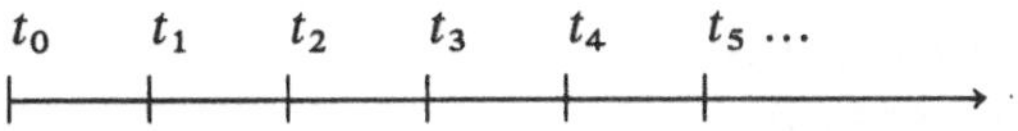

Consider now a proposition A. In order to describe the possible variety of the truth values of A at different times t the simplest linguistic means would be to introduce an explicit time parameter in the proposition and denote it by $A(t)$. The main idea of the logical language we want to define, however, is to avoid just such an explicit occurrence of time. The reason is that we do not really want to be able to express assertions like "A is true at t_{17}". We rather want to have at hand logical *operators* which enable us to formulate new propositions about the truth values of A at time points which are related to some reference point (the *present* time) in particular ways.

Examples of desirable operators with their intended (informal) meaning are:

$\circ A$: "A holds at the time point immediately after the reference point" (*nexttime* operator).

$\Box A$: "A holds at all time points after the reference point" (*always* or *henceforth* operator).

$\Diamond A$: "There is a time point after the reference point at which A holds" (*sometime* or *eventually* operator).

A **atnext** B: "A will hold at the next time point that B holds" (*first time* or *atnext* operator).

A **until** B: "A holds at all following time points up to a time point at which B holds" (*until* operator).

Of course, the truth of such a new formula is itself time dependent, because it may differ with different "reference points". If, for example,

A is true at t_4 and
A is false at t_5

then:

$\circ A$ is true at t_3 and false at t_4.

But formally this time dependence will only occur as a semantical concept in order to define some notion of validity for such formulas. It does not occur in the language itself.

Observe finally that all of the operators "look into the future". In fact we are not interested in operators which relate the reference point to former time points. At a first glance this unsymmetry may seem unnatural, but it will turn out that our operators will be quite appropriate for our purposes.

Even at this informal level of discussion we see that the *temporal operators* we want to introduce are *propositional* in the sense that they do not refer to subject variables. Hence it is reasonable to develop the main concepts of this temporal logic in the framework of propositional logic. So we now define formally:

A *language* $\mathscr{L}_{\mathrm{TA}}$ *of propositional temporal logic* is given as follows:

Alphabet

- A denumerable set $\mathscr{V}$ of *atomic formulas*,
- the symbols $\neg$, $\rightarrow$, $\circ$, $\Box$, **atnext**, (,).

Inductive definition of *formulas*:

1. Every atomic formula is a formula.
2. If A is a formula then $\neg A$, $\circ A$ and $\Box A$ are formulas.
3. If A and B are formulas then $(A \rightarrow B)$ and $(A \ \mathbf{atnext}\ B)$ are formulas.

Further operators can be introduced as abbreviations, e.g.,

$\wedge$, $\vee$, $\leftrightarrow$, **true**, **false** as in classical logic,
$\Diamond A$ for $\neg\Box\neg A$.

We also introduce the *iterated* atnext operator $\mathbf{atnext}^n$ inductively defined as follows:

$$(A\ \mathbf{atnext}^1\ B) \quad \text{is} \quad (A\ \mathbf{atnext}\ B),$$
$$(A\ \mathbf{atnext}^{n+1}\ B) \quad \text{is} \quad ((A\ \mathbf{atnext}^n\ B)\ \mathbf{atnext}\ B).$$

So, e.g., $(A\ \mathbf{atnext}^2\ B)$ is the same as $((A\ \mathbf{atnext}\ B)\ \mathbf{atnext}\ B)$.

In the following, we will use the *syntactic variables*:

$v, v_1, v_2, \ldots$	for atomic formulas,
$A, A_1, A_2, B, B_1, \ldots, F, \ldots$	for formulas,

and for notational simplicity we establish a priority order of the operators:

$\neg$, $\circ$, $\Box$, $\Diamond$	have higher priority than all binary operators,
atnext	has higher priority than $\wedge$, $\vee$, $\rightarrow$, $\leftrightarrow$,
$\wedge$, $\vee$	have higher priority than $\rightarrow$, $\leftrightarrow$,
$\rightarrow$	has higher priority than $\leftrightarrow$.

Accordingly, we will omit superfluous parentheses (including the outermost).

Example. Instead of:

$$((\circ A_1 \vee A_2) \to (\neg A_3 \wedge (\Box A_4 \ \mathtt{atnext} \ A_5)))$$

we write:

$$\circ A_1 \vee A_2 \to \neg A_3 \wedge \Box A_4 \ \mathtt{atnext} \ A_5. \quad \Box$$

2. Semantics of $\mathscr{L}_{TA}$

The basic semantical notion of classical propositional logic is that of a valuation. For a language $\mathscr{L}_{TA}$ of propositional temporal logic we have to extend this concept according to our informal idea that formulas are "valuated" over a time scale.

A *temporal* (or *Kripke*) *structure* **K** for $\mathscr{L}_{TA}$ consists of

- an infinite sequence $\{\eta_0, \eta_1, \eta_2, \ldots\}$ of mappings

  $$\eta_i : \mathscr{V} \to \{\mathbf{f}, \mathbf{t}\};$$

  the η_i are called *states*. η_0 is the *initial state*.

The infinite sequence of states formalizes the informal time scale; a state is a "time point". Every state is a valuation in the classical sense.

For every temporal structure **K**, every $i \in \mathbb{N}_0$ and every formula F we inductively define the truth value $\mathbf{K}_i(F) \in \{\mathbf{f}, \mathbf{t}\}$, informally meaning the "truth value of F in state η_i":

1. $\mathbf{K}_i(v) = \eta_i(v)$ for $v \in \mathscr{V}$.
2. $\mathbf{K}_i(\neg A) = \mathbf{t}$ iff $\mathbf{K}_i(A) = \mathbf{f}$.
3. $\mathbf{K}_i(A \to B) = \mathbf{t}$ iff $\mathbf{K}_i(A) = \mathbf{f}$ or $\mathbf{K}_i(B) = \mathbf{t}$.
4. $\mathbf{K}_i(\circ A) = \mathbf{t}$ iff $\mathbf{K}_{i+1}(A) = \mathbf{t}$.
5. $\mathbf{K}_i(\Box A) = \mathbf{t}$ iff $\mathbf{K}_j(A) = \mathbf{t}$ for every $j \geq i$.
6. $\mathbf{K}_i(A \ \mathtt{atnext} \ B) = \mathbf{t}$ iff $\mathbf{K}_j(B) = \mathbf{f}$ for every $j > i$ or
   $\mathbf{K}_k(A) = \mathbf{t}$ for the smallest $k > i$ with
   $\mathbf{K}_k(B) = \mathbf{t}$.

Obviously, the operators $\neg$ and $\to$ are the old classical operators (in every state) without any temporal aspect. The definitions for $\circ$, $\Box$ and **atnext** just formalize our informal intentions given in Section 1. Notice, however, that in the definition for the operator $\Box$ we "include the present" by stating $\mathbf{K}_j(A) = \mathbf{t}$ for every $j \geq i$ (and not only $j > i$), and that in the definition for **atnext** we do not claim that $\mathbf{K}_k(B) = \mathbf{t}$ for some $k > i$ (informally: B need not become true).

Example. Let $A \equiv \circ \neg v_1 \ \mathtt{atnext} \ \Box v_2$, and let **K** be according to the following matrix:

	η_0	η_1	η_2	η_3	η_4	η_5	η_6	η_7	η_8	...
v_1:	f	f	f	f	f	t	f	t	f	...~...
v_2:	t	f	f	t	f	t	t	t	t	...⟨t forever⟩... .

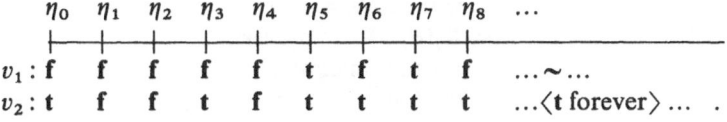

The two lines represent the values of v_1 and v_2, respectively, for η_0, η_1, η_2, etc. The $\sim$ sign indicates arbitrary values. It is easy to compute that:

$$\mathbf{K}_0(A)=\mathbf{K}_1(A)=\mathbf{K}_2(A)=\mathbf{K}_3(A)=\mathbf{K}_4(A)=\mathbf{t}$$

since in every case, the smallest $k>i$ ($i=1, \ldots, 4$) with $\mathbf{K}_k(\Box v_2)=\mathbf{t}$ is $k=5$ and $\mathbf{K}_5(\bigcirc\neg v_1)=\mathbf{t}$ since $\mathbf{K}_6(v_1)=\mathbf{f}$. In the same way we find:

$\mathbf{K}_5(A)=\mathbf{f}$ (because of $\mathbf{K}_7(v_1)=\mathbf{t}$), and
$\mathbf{K}_6(A)=\mathbf{t}$ (because of $\mathbf{K}_8(v_1)=\mathbf{f}$). □

Of course, the definitions 1–6 above also imply truth values for formulas with the other operators:

$\mathbf{K}_i(A \wedge B) = \mathbf{t}$ iff $\mathbf{K}_i(A)=\mathbf{t}$ and $\mathbf{K}_i(B)=\mathbf{t}$.
$\mathbf{K}_i(A \vee B) = \mathbf{t}$ iff $\mathbf{K}_i(A)=\mathbf{t}$ or $\mathbf{K}_i(B)=\mathbf{t}$.
$\mathbf{K}_i(A \leftrightarrow B) = \mathbf{t}$ iff $\mathbf{K}_i(A)=\mathbf{K}_i(B)$.
$\mathbf{K}_i(\mathtt{true}) = \mathbf{t}$.
$\mathbf{K}_i(\mathtt{false}) = \mathbf{f}$.
$\mathbf{K}_i(\Diamond A) = \mathbf{t}$ iff $\mathbf{K}_j(A)=\mathbf{t}$ for some $j \geq i$.
$\mathbf{K}_i(A\ \mathtt{atnext}^2\ B)=\mathbf{t}$ iff $\mathbf{K}_j(B)=\mathbf{t}$ for at most one $j>i$ or
$\mathbf{K}_k(A)=\mathbf{t}$ for the second smallest
$k>i$ with $\mathbf{K}_k(B)=\mathbf{t}$.

The latter definition expresses informally the phrase

"A holds in the second next state in which B holds".

In general, $A\ \mathtt{atnext}^n\ B$ means:

"A holds in the n-th next state in which B holds",

formally given by:

$\mathbf{K}_i(A\ \mathtt{atnext}^n\ B)=\mathbf{t}$ iff $\mathbf{K}_j(B)=\mathbf{t}$ for at most $n-1$ $j>i$ or
$\mathbf{K}_k(A)=\mathbf{t}$ for the n-th smallest $k>i$ with
$\mathbf{K}_k(B)=\mathbf{t}$.

We only give proofs for $\Diamond$ and $\mathtt{atnext}^2$:

$\mathbf{K}_i(\Diamond A)=\mathbf{t} \Leftrightarrow \mathbf{K}_i(\neg\Box\neg A)=\mathbf{t}$
$\Leftrightarrow \mathbf{K}_i(\Box\neg A) = \mathbf{f}$
$\Leftrightarrow \mathbf{K}_j(\neg A) = \mathbf{f}$ for some $j \geq i$
$\Leftrightarrow \mathbf{K}_j(A) = \mathbf{t}$ for some $j \geq i$.

$\mathbf{K}_i(A\ \mathtt{atnext}^2\ B)=\mathbf{t} \Leftrightarrow \mathbf{K}_j(B)=\mathbf{f}$ for every $j>i$ or
$\mathbf{K}_k(A\ \mathtt{atnext}\ B)=\mathbf{t}$ for the smallest $k>i$
with $\mathbf{K}_k(B)=\mathbf{t}$
$\Leftrightarrow \mathbf{K}_j(B)=\mathbf{f}$ for every $j>i$ or
for the smallest $k>i$ with $\mathbf{K}_k(B)=\mathbf{t}$
the following holds:

$$\begin{aligned}
&\mathbf{K}_{j'}(B)=\mathbf{f} \quad \text{for every } j'>k \quad \text{or}\\
&\mathbf{K}_{k'}(A)=\mathbf{t} \quad \text{for the smallest } k'>k\\
&\quad \text{with } \mathbf{K}_{k'}(B)=\mathbf{t}\\
\Leftrightarrow\ &\mathbf{K}_j(B)=\mathbf{t} \quad \text{for at most one } j>i \quad \text{or}\\
&\mathbf{K}_k(A)=\mathbf{t} \quad \text{for the second smallest } k>i \quad \text{with}\\
&\mathbf{K}_k(B)=\mathbf{t}. \quad \square
\end{aligned}$$

Here (and in the following) we use the symbol $\Leftrightarrow$ for the (*metalogical*) equivalence of assertions. Later we will also use $\Rightarrow$ for metalogical implication.

In the definition of $\mathscr{L}_{TA}$ the operator $\Diamond$ was not introduced as a "basic" operator but defined via $\Box$ (and $\neg$). It was not even necessary to introduce $\circ$ and $\Box$ as basic elements; both are expressible by `atnext` because of the following facts:

$$\mathbf{K}_i(\circ A)=\mathbf{K}_i(A\ \mathtt{atnext}\ \mathtt{true}).$$
$$\mathbf{K}_i(\Box A)=\mathbf{K}_i(A\wedge \mathtt{false}\ \mathtt{atnext}\ \neg A).$$

Proof

$$\begin{aligned}
\mathbf{K}_i(A\ \mathtt{atnext}\ \mathtt{true})=\mathbf{t}\Leftrightarrow\ &\mathbf{K}_j(\mathtt{true})=\mathbf{f} \quad \text{for every } j>i \quad \text{or}\\
&\mathbf{K}_k(A)=\mathbf{t} \quad \text{for the smallest } k>i \quad \text{with}\\
&\mathbf{K}_k(\mathtt{true})=\mathbf{t}\\
\Leftrightarrow\ &\mathbf{K}_{i+1}(A)=\mathbf{t}\\
\Leftrightarrow\ &\mathbf{K}_i(\circ A)=\mathbf{t}.
\end{aligned}$$

$$\begin{aligned}
\mathbf{K}_i(A\wedge \mathtt{false}\ \mathtt{atnext}\ \neg A)=\mathbf{t}\Leftrightarrow\ &\mathbf{K}_i(A)=\mathbf{t} \quad \text{and}\\
&\mathbf{K}_i(\mathtt{false}\ \mathtt{atnext}\ \neg A)=\mathbf{t}\\
\Leftrightarrow\ &\mathbf{K}_i(A)=\mathbf{t} \quad \text{and}\\
&[\mathbf{K}_j(\neg A)=\mathbf{f} \quad \text{for every } j>i \quad \text{or}\\
&\mathbf{K}_k(\mathtt{false})=\mathbf{t} \quad \text{for the smallest}\\
&k>i \quad \text{with } \mathbf{K}_k(\neg A)=\mathbf{t}]\\
\Leftrightarrow\ &\mathbf{K}_i(A)=\mathbf{t} \quad \text{and} \quad \mathbf{K}_j(\neg A)=\mathbf{f}\\
&\text{for every } j>i\\
\Leftrightarrow\ &\mathbf{K}_j(A)=\mathbf{t} \quad \text{for every } j\geq i\\
\Leftrightarrow\ &\mathbf{K}_i(\Box A)=\mathbf{t}. \quad \square
\end{aligned}$$

Thus, $\circ A$ and $\Box A$ could be introduced for

A `atnext true`,

$A\wedge$ `false atnext` $\neg A$, respectively.

The reason for taking $\circ$ and $\Box$ as basic symbols in $\mathscr{L}_{TA}$ is only technical convenience.

The concept of temporal structures is a generalization of the concept of valuations. It also yields the notion of validity.

Definition. A formula A is called *valid in the temporal structure* $\mathbf{K}$ (denoted by $\Vdash_{\mathbf{K}} A$) if $\mathbf{K}_i(A)=\mathbf{t}$ for every $i\in\mathbb{N}_0$. A is called *valid* (denoted by $\Vdash A$) if $\Vdash_{\mathbf{K}} A$ for every $\mathbf{K}$. *A follows from* a set $\mathscr{F}$ of formulas (denoted by $\mathscr{F}\Vdash A$) if $\Vdash_{\mathbf{K}} A$ for every $\mathbf{K}$ with $\Vdash_{\mathbf{K}} B$ for every $B\in\mathscr{F}$.

We now state some fundamental facts about these notions.

Theorem 2.1. *If $\mathscr{F}\Vdash A$ and $\Vdash B$ for every $B\in\mathscr{F}$ then $\Vdash A$.*

Proof. Let $\mathbf{K}$ be a temporal structure. Then $\Vdash_{\mathbf{K}} B$ for every $B \in \mathscr{F}$ and hence $\Vdash_{\mathbf{K}} A$. Since this holds for every $\mathbf{K}$, we have $\Vdash A$. □

Theorem 2.2. *$A_1, \ldots, A_n \Vdash B$ if and only if $\Vdash \Box A_1 \wedge \ldots \wedge \Box A_n \rightarrow B$.*

Proof. We first show the direction "⇒":
Let $A_1, \ldots, A_n \Vdash B$, $\mathbf{K} = \{\eta_0, \eta_1, \eta_2, \ldots\}$ and $i \in \mathbb{N}_0$. Assume $\mathbf{K}_i(\Box A_1 \wedge \ldots \wedge \Box A_n \rightarrow B) = \mathbf{f}$. This means $\mathbf{K}_i(\Box A_1) = \ldots = \mathbf{K}_i(\Box A_n) = \mathbf{t}$ and $\mathbf{K}_i(B) = \mathbf{f}$ and hence $\mathbf{K}_j(A_1) = \ldots = \mathbf{K}_j(A_n) = \mathbf{t}$ for every $j \geq i$ and $\mathbf{K}_i(B) = \mathbf{f}$. Let now $\mathbf{K}' = \{\eta'_0, \eta'_1, \eta'_2, \ldots\}$ be a new temporal structure with $\eta'_j = \eta_{i+j}$ for every $j \in \mathbb{N}_0$. Then we have $\mathbf{K}'_j(A_1) = \ldots = \mathbf{K}'_j(A_n) = \mathbf{t}$ for every $j \in \mathbb{N}_0$ and $\mathbf{K}'_0(B) = \mathbf{f}$, hence $\Vdash_{\mathbf{K}'} A_1, \ldots, \Vdash_{\mathbf{K}'} A_n$ but not $\Vdash_{\mathbf{K}'} B$. This is a contradiction, so $\mathbf{K}_i(\Box A_1 \wedge \ldots \wedge \Box A_n \rightarrow B) = \mathbf{t}$ and since $\mathbf{K}$ and i are arbitrary, we get $\Vdash \Box A_1 \wedge \ldots \wedge \Box A_n \rightarrow B$.

The opposite direction "⇐":
Let $\Vdash \Box A_1 \wedge \ldots \wedge \Box A_n \rightarrow B$, $\mathbf{K}$ be a temporal structure such that $\Vdash_{\mathbf{K}} A_1, \ldots, \Vdash_{\mathbf{K}} A_n$ and $i \in \mathbb{N}_0$. Then $\mathbf{K}_j(A_1) = \ldots = \mathbf{K}_j(A_n) = \mathbf{t}$ for every $j \in \mathbb{N}_0$, and hence also $\mathbf{K}_j(A_1) = \ldots = \mathbf{K}_j(A_n) = \mathbf{t}$ for every $j \geq i$. This implies that $\mathbf{K}_i(\Box A_1 \wedge \ldots \wedge \Box A_n) = \mathbf{t}$ and because of $\mathbf{K}_i(\Box A_1 \wedge \ldots \wedge \Box A_n \rightarrow B) = \mathbf{t}$ we also get $\mathbf{K}_i(B) = \mathbf{t}$. This means $\Vdash_{\mathbf{K}} B$ and we have the desired result that $A_1, \ldots, A_n \Vdash B$. □

Theorem 2.2 is the temporal logic analogon of the classical fact:

$$A_1, \ldots, A_n \Vdash B \quad \text{iff} \quad \Vdash A_1 \wedge \ldots \wedge A_n \rightarrow B.$$

It should be noted that the latter no longer holds in $\mathscr{L}_{\mathrm{TA}}$. A simple counterexample is given by $A \Vdash \Box A$. This holds but $A \rightarrow \Box A$ is not valid.

Theorem 2.3. *If $\mathscr{F} \Vdash A$ and $\mathscr{F} \Vdash A \rightarrow B$ then $\mathscr{F} \Vdash B$.*

Proof. Let $\mathbf{K}$ be a temporal structure such that $\Vdash_{\mathbf{K}} C$ for every $C \in \mathscr{F}$, and $i \in \mathbb{N}_0$. Then $\mathbf{K}_i(A) = \mathbf{K}_i(A \rightarrow B) = \mathbf{t}$ and hence $\mathbf{K}_i(B) = \mathbf{t}$. This means that $\mathscr{F} \Vdash B$. □

Theorem 2.3 states that *modus ponens* is a "valid rule of inference".

Theorem 2.4. *If $\mathscr{F} \Vdash A$ then $\mathscr{F} \Vdash \circ A$ and $\mathscr{F} \Vdash \Box A$.*

Proof. Let $\mathbf{K}$ be a temporal structure such that $\Vdash_{\mathbf{K}} C$ for every $C \in \mathscr{F}$, and $i \in \mathbb{N}_0$. Then $\mathbf{K}_j(A) = \mathbf{t}$ for every $j \in \mathbb{N}_0$, in particular $\mathbf{K}_{i+1}(A) = \mathbf{t}$ and $\mathbf{K}_j(A) = \mathbf{t}$ for every $j \geq i$. This means that $\mathscr{F} \Vdash \circ A$ and $\mathscr{F} \Vdash \Box A$. □

We define yet another semantic notion which will be needed in Section 6.

Definition. A set $\mathscr{F}$ of formulas is called *satisfiable* if there is some temporal structure $\mathbf{K}$ and $i \in \mathbb{N}_0$ such that $\mathbf{K}_i(A) = \mathbf{t}$ for every $A \in \mathscr{F}$. A formula A is called satisfiable if $\{A\}$ is satisfiable.

Example. Let $\mathscr{F} = \{v_1 \wedge \Box v_2 \wedge v_3 \ \mathbf{atnext} \ \neg v_1, (v_2 \rightarrow \circ v_1) \vee \circ \neg v_2\}$. Take the following $\mathbf{K}$:

	η_0	η_1	η_2	η_3	...
v_1 :	**t**	**t**	**f**	~	...
v_2 :	**t**	**t**	**t**	**t**	... ⟨**t** forever⟩...
v_3 :	~	~	**t**	~	....

It can easily be computed that:

$$\mathbf{K}_0(v_1 \wedge \Box v_2 \wedge v_3 \ \mathbf{atnext} \ \neg v_1) = \mathbf{t}, \quad \text{and}$$
$$\mathbf{K}_0((v_2 \rightarrow \circ v_1) \vee \circ \neg v_2) = \mathbf{t}.$$

Hence, $\mathscr{F}$ is satisfiable. □

Validity and satisfiability are "dual" notions in the following sense:

Theorem 2.5. *$\Vdash A$ if and only if $\neg A$ is not satisfiable.*

Proof. $\Vdash A$ iff $\mathbf{K}_i(A) = \mathbf{t}$ and hence $\mathbf{K}_i(\neg A) = \mathbf{f}$ for every $\mathbf{K}$ and every $i \in \mathbb{N}_0$. This just holds iff $\neg A$ is not satisfiable. □

We conclude this section by listing some more temporal formulas with their informal meanings. We recommend that the reader check these informal formulations very carefully against the formal semantics.

$A \rightarrow \Box B$:	"If A then henceforth B",
$A \rightarrow \Diamond B$:	"If A then sometime (now or in the future) B",
$\Box(A \rightarrow B)$:	"Whenever henceforth A then B",
$\Diamond(A \wedge \circ \neg A)$:	"Sometime A and immediately after that not A",
$\Diamond \Box A$:	"Sometime A will hold permanently",
$\Box \Diamond A$:	"For every following state there is a later state where A holds", i.e., "A holds infinitely often from now on",
$A \wedge B \rightarrow A \ \mathbf{atnext} \ B$:	"If A and B hold then A also holds next time when B holds".

3. Temporal Logical Laws

In any logic, the valid formulas express "logical laws". An example from classical logic is *de Morgan's law*

$$\neg(A \wedge B) \leftrightarrow \neg A \vee \neg B$$

stating "duality" of the operators $\wedge$ and $\vee$. According to our semantical definitions in the previous section we should expect that such tautologies remain valid also in temporal logic where we may substitute formulas of $\mathscr{L}_{\text{TA}}$ for A and B, e.g., in the law above:

$$\neg(\circ C \wedge \Box D) \leftrightarrow \neg \circ C \vee \neg \Box D.$$

Let us formally distinguish the set of such formulas.

Definition. A formula (of $\mathscr{L}_{TA}$) is called *tautologically valid* if it results from a tautology A (of $\mathscr{L}_A$) by consistently replacing the atomic formulas of A by formulas of $\mathscr{L}_{TA}$.

Our expectation noted above is confirmed by the following theorem.

Theorem 3.1. *Every tautologically valid formula is valid.*

Proof. Let A^* result from a formula A of $\mathscr{L}_A$ by replacing the atomic formulas $v_1, \dots, v_n$ by formulas $A_1, \dots, A_n$ of $\mathscr{L}_{TA}$ and let $\mathbf{K}$ be a temporal structure and $i \in \mathbb{N}_0$. Define a classical valuation $\mathbf{B}$ of $\mathscr{L}_A$ by $\mathbf{B}(v_j) = \mathbf{K}_i(A_j)$ for $j = 1, \dots, n$ (and with arbitrary values for other $v_k \in \mathscr{V}$). We claim that:

(*) $\mathbf{B}(A) = \mathbf{K}_i(A^*)$

which proves the theorem, since $\mathbf{B}(A) = \mathbf{t}$ if A is a tautology. The proof of (*) runs by induction on (the syntax of) A:

1. $A \equiv v_j$: then $A^* \equiv A_j$, hence $\mathbf{B}(A) = \mathbf{B}(v_j) = \mathbf{K}_i(A_j) = \mathbf{K}_i(A^*)$.
2. $A \equiv \neg B$: with the corresponding meaning of the "operator" * on B we have $A^* \equiv \neg B^*$, hence using the induction hypothesis

$$\mathbf{B}(A) = \mathbf{t} \Leftrightarrow \mathbf{B}(B) = \mathbf{f} \Leftrightarrow \mathbf{K}_i(B^*) = \mathbf{f} \Leftrightarrow \mathbf{K}_i(A^*) = \mathbf{t}.$$

3. $A \equiv B_1 \to B_2$: we have again $A^* \equiv B_1^* \to B_2^*$ and with the induction hypothesis we get:

$$\begin{aligned} \mathbf{B}(A) = \mathbf{t} &\Leftrightarrow \mathbf{B}(B_1) = \mathbf{f} \quad \text{or} \quad \mathbf{B}(B_2) = \mathbf{t} \\ &\Leftrightarrow \mathbf{K}_i(B_1^*) = \mathbf{f} \quad \text{or} \quad \mathbf{K}_i(B_2^*) = \mathbf{t} \\ &\Leftrightarrow \mathbf{K}_i(B_1^* \to B_2^*) = \mathbf{t}. \quad \square \end{aligned}$$

According to this theorem we may use all tautologies of classical logic as "logical laws" in temporal logic. We also want to interpret this result in a somewhat different way. Suppose a formula B follows from some formulas $A_1, \dots, A_n$ in classical propositional logic. Again, if we substitute (consistently) formulas of $\mathscr{L}_{TA}$ in $A_1, \dots, A_n$ and B, we should not destroy the logical relationship. For example, we should have:

$$\circ A \to \Box B,\ \Box B \to \Diamond C \Vdash \circ A \to \Diamond C$$

since in classical logic we have:

$$A' \to B',\ B' \to C' \Vdash A' \to C'.$$

For a simple formulation of this fact we remember that in classical propositional logic we have:

$$A_1, \dots, A_n \Vdash B \quad \text{iff} \quad \Vdash A_1 \wedge \dots \wedge A_n \to B$$

and so we may define:

Definition. Let $A_1, \dots, A_n, B$ $(n \geq 1)$ be formulas (of $\mathscr{L}_{TA}$). B is called a *tautological consequence* of $A_1, \dots, A_n$ if the formula $A_1 \wedge \dots \wedge A_n \to B$ is tautologically valid.

The desired result is:

Theorem 3.2. *If B is a tautological consequence of $A_1, \ldots, A_n$, then $A_1, \ldots, A_n \Vdash B$.*

Proof. Let $\mathbf{K}$ be a temporal structure with $\Vdash_{\mathbf{K}} A_j$, $j=1, \ldots, n$ and $i \in \mathbb{N}_0$. Then we have $\mathbf{K}_i(A_1) = \ldots = \mathbf{K}_i(A_n) = \mathbf{t}$, i.e., $\mathbf{K}_i(A_1 \wedge \ldots \wedge A_n) = \mathbf{t}$, and $\mathbf{K}_i(A_1 \wedge \ldots \wedge A_n \rightarrow B) = \mathbf{t}$. This implies $\mathbf{K}_i(B) = \mathbf{t}$ and proves the theorem. □

So far we know of logical laws in $\mathscr{L}_{\mathrm{TA}}$ coming from the "classical part" of the new logic. Let us now turn to proper temporal logical laws concerning the temporal operators. We give quite an extensive list of formulas all of which we claim to be valid without proving these facts except for a few examples.

Duality laws

(T1) $\quad \neg \circ A \leftrightarrow \circ \neg A$

(T2) $\quad \neg \Box A \leftrightarrow \Diamond \neg A$

(T3) $\quad \neg \Diamond A \leftrightarrow \Box \neg A$

(T2) and (T3) describe the *duality* of $\Box$ and $\Diamond$. (T1) expresses that $\circ$ is *self-dual.*

Proof of (T1)

$$\begin{aligned} \mathbf{K}_i(\neg \circ A) = \mathbf{t} &\Leftrightarrow \mathbf{K}_i(\circ A) = \mathbf{f} \\ &\Leftrightarrow \mathbf{K}_{i+1}(A) = \mathbf{f} \\ &\Leftrightarrow \mathbf{K}_{i+1}(\neg A) = \mathbf{t} \\ &\Leftrightarrow \mathbf{K}_i(\circ \neg A) = \mathbf{t}. \quad \Box \end{aligned}$$

Reflexivity laws

(T4) $\quad \Box A \rightarrow A$

(T5) $\quad A \rightarrow \Diamond A$

These formulas express the fact that (with respect to $\Box$ and $\Diamond$) the "future" includes the "present".

Laws about the "strength" of the operators

(T6) $\quad \Box A \rightarrow \circ A$

(T7) $\quad \circ A \rightarrow \Diamond A$

(T8) $\quad \Box A \rightarrow \Diamond A$

(T9) $\quad \Box A \rightarrow A \ \mathtt{atnext}\ B$

(T10) $\quad \Diamond \Box A \rightarrow \Box \Diamond A$

Proof of (T10)

$$\begin{aligned} \mathbf{K}_i(\Diamond \Box A) = \mathbf{t} &\Rightarrow \mathbf{K}_j(\Box A) = \mathbf{t} && \text{for some } j \geq i \\ &\Rightarrow \mathbf{K}_k(A) = \mathbf{t} && \text{for every } k \geq j \quad \text{and some } j \geq i \\ &\Rightarrow [\mathbf{K}_k(A) = \mathbf{t} && \text{for some } k \geq j] \\ & && \text{for every } j \geq i \\ &\Rightarrow \mathbf{K}_j(\Diamond A) = \mathbf{t} && \text{for every } j \geq i \\ &\Rightarrow \mathbf{K}_i(\Box \Diamond A) = \mathbf{t}. && \Box \end{aligned}$$

Expressibility laws

(T11) $\circ A \leftrightarrow A$ **atnext true**

(T12) $\Box A \leftrightarrow A \wedge$ **false atnext** $\neg A$

(T13) $\Diamond A \leftrightarrow A \vee \neg($**false atnext** $A)$

These formulas state the expressibility of $\circ$, $\Box$, $\Diamond$ by **atnext**. (T11) and (T12) were proved in the previous section.

Idempotency laws

(T14) $\Box\Box A \leftrightarrow \Box A$

(T15) $\Diamond\Diamond A \leftrightarrow \Diamond A$

Proof of (T14)

$$\begin{aligned} \mathbf{K}_i(\Box\Box A) = \Leftrightarrow & \mathbf{K}_j(\Box A) = \mathbf{t} && \text{for every } j \geq i \\ \Leftrightarrow & \mathbf{K}_k(A) = \mathbf{t} && \text{for every } k \geq j \text{ and every } j \geq i \\ \Leftrightarrow & \mathbf{K}_k(A) = \mathbf{t} && \text{for every } k \geq i \\ \Leftrightarrow & \mathbf{K}_i(\Box A) = \mathbf{t}. && \Box \end{aligned}$$

Commutativity laws

(T16) $\Box \circ A \leftrightarrow \circ \Box A$

(T17) $\Diamond \circ A \leftrightarrow \circ \Diamond A$

These formulas state the *commutativity* of $\circ$ with $\Box$ and $\Diamond$.

Proof of (T16)

$$\begin{aligned} \mathbf{K}_i(\Box\circ A) = \mathbf{t} \Leftrightarrow & \mathbf{K}_j(\circ A) = \mathbf{t} && \text{for every } j \geq i \\ \Leftrightarrow & \mathbf{K}_{j+1}(A) = \mathbf{t} && \text{for every } j \geq i \\ \Leftrightarrow & \mathbf{K}_j(A) = \mathbf{t} && \text{for every } j \geq i+1 \\ \Leftrightarrow & \mathbf{K}_{i+1}(\Box A) = \mathbf{t} && \\ \Leftrightarrow & \mathbf{K}_i(\circ\Box A) = \mathbf{t}. && \Box \end{aligned}$$

Distributivity laws

(T18) $\circ(A \rightarrow B) \leftrightarrow \circ A \rightarrow \circ B$

(T19) $\circ(A \wedge B) \leftrightarrow \circ A \wedge \circ B$

(T20) $\circ(A \vee B) \leftrightarrow \circ A \vee \circ B$

(T21) $\circ(A$ **atnext** $B) \leftrightarrow \circ A$ **atnext** $\circ B$

(T22) $\Box(A \wedge B) \leftrightarrow \Box A \wedge \Box B$

(T23) $\Diamond(A \vee B) \leftrightarrow \Diamond A \vee \Diamond B$

(T24) $(A \wedge B)$ **atnext** $C \leftrightarrow A$ **atnext** $C \wedge B$ **atnext** C

(T25) $(A \vee B)$ **atnext** $C \leftrightarrow A$ **atnext** $C \vee B$ **atnext** C

(T18)–(T21) express the *distributivity* of $\circ$ over all binary operators (we have not explicitly listed the distributivity over $\leftrightarrow$ but this follows, of course, from (T18) and (T19)). The other formulas state distributivity laws of $\Box$, $\Diamond$, and **atnext**.

Proof of (T18)

$$\begin{aligned}\mathbf{K}_i(\circ(A\to B))=\mathbf{t} &\Leftrightarrow \mathbf{K}_{i+1}(A\to B)=\mathbf{t}\\ &\Leftrightarrow \mathbf{K}_{i+1}(A)=\mathbf{f} \quad\text{or}\quad \mathbf{K}_{i+1}(B)=\mathbf{t}\\ &\Leftrightarrow \mathbf{K}_i(\circ A)=\mathbf{f} \quad\text{or}\quad \mathbf{K}_i(\circ B)=\mathbf{t}\\ &\Leftrightarrow \mathbf{K}_i(\circ A\to\circ B)=\mathbf{t}. \quad \square\end{aligned}$$

Weak distributivity laws

(T26) $\square(A\to B)\to(\square A\to\square B)$

(T27) $\square A\vee\square B\to\square(A\vee B)$

(T28) $(\Diamond A\to\Diamond B)\to\Diamond(A\to B)$

(T29) $\Diamond(A\wedge B)\to\Diamond A\wedge\Diamond B$

(T30) $A\,\mathtt{atnext}(B\vee C)\to A\;\mathtt{atnext}\;B\vee A\;\mathtt{atnext}\;C$

These formulas state that at least "some direction" of further distributivities of $\square$, $\Diamond$, and **atnext** hold.

Proof of (T27)

$$\begin{aligned}\mathbf{K}_i(\square A\vee\square B)=\mathbf{t} &\Rightarrow \mathbf{K}_i(\square A)=\mathbf{t} \quad\text{or}\quad \mathbf{K}_i(\square B)=\mathbf{t}\\ &\Rightarrow \mathbf{K}_j(A)=\mathbf{t} \quad\text{for every } j\ge i \quad\text{or}\\ &\quad\;\; \mathbf{K}_j(B)=\mathbf{t} \quad\text{for every } j\ge i\\ &\Rightarrow \mathbf{K}_j(A)=\mathbf{t} \quad\text{or}\quad \mathbf{K}_j(B)=\mathbf{t} \quad\text{for every } j\ge i\\ &\Rightarrow \mathbf{K}_j(A\vee B)=\mathbf{t} \quad\text{for every } j\ge i\\ &\Rightarrow \mathbf{K}_i(\square(A\vee B))=\mathbf{t}. \quad \square\end{aligned}$$

Recursion equivalences

(T31) $\square A\leftrightarrow A\wedge\circ\square A$

(T32) $\Diamond A\leftrightarrow A\vee\circ\Diamond A$

(T33) $A\;\mathtt{atnext}\;B\leftrightarrow\circ(B\to A)\wedge\circ(\neg B\to A\;\mathtt{atnext}\;B)$

(T31) is a kind of "recursive" formulation of the informal representation:

$$\square A\leftrightarrow A\wedge\circ A\wedge\circ\circ A\wedge\circ\circ\circ A\wedge\ldots$$

of a formula $\square A$. (T32) and (T33) are analogous for $\Diamond$ and **atnext**

Proof of (T31)

$$\begin{aligned}\mathbf{K}_i(A\wedge\circ\square A)=\mathbf{t} &\Leftrightarrow \mathbf{K}_i(A)=\mathbf{t} \quad\text{and}\quad \mathbf{K}_i(\circ\square A)=t\\ &\Leftrightarrow \mathbf{K}_i(A)=\mathbf{t} \quad\text{and}\quad \mathbf{K}_j(A)=\mathbf{t} \quad\text{for every } j\ge i+1\\ &\Leftrightarrow \mathbf{K}_j(A)=\mathbf{t} \quad\text{for every } j\ge i\\ &\Leftrightarrow \mathbf{K}_j(\square A)=\mathbf{t}. \quad \square\end{aligned}$$

Proof of (T33)

$$\begin{aligned}&\mathbf{K}_i(\circ(B\to A)\wedge\circ(\neg B\to A\;\mathtt{atnext}\;B))=\mathbf{t}\\ &\quad\Leftrightarrow [\mathbf{K}_{i+1}(B)=\mathbf{f} \quad\text{or}\quad \mathbf{K}_{i+1}(A)=\mathbf{t}] \quad\text{and}\\ &\quad\quad\;\; [\mathbf{K}_{i+1}(\neg B)=\mathbf{f} \quad\text{or}\quad \mathbf{K}_{i+1}(A\;\mathtt{atnext}\;B)=\mathbf{t}]\end{aligned}$$

$\Leftrightarrow \mathbf{K}_{i+1}(A)=\mathbf{K}_{i+1}(B)=\mathbf{t}$ or
$[\mathbf{K}_{i+1}(B)=\mathbf{f}$ and $\mathbf{K}_{i+1}(A\ \mathtt{atnext}\ B)=\mathbf{t}]$
$\Leftrightarrow \mathbf{K}_{i+1}(A)=\mathbf{K}_{i+1}(B)=\mathbf{t}$ or
$[\mathbf{K}_j(B)=\mathbf{f}$ for every $j>i$ or
$[\mathbf{K}_{i+1}(B)=\mathbf{f}$ and $\mathbf{K}_k(A)=\mathbf{t}$
for the smallest $k>i+1$ with $\mathbf{K}_k(B)=\mathbf{t}]]$
$\Leftrightarrow \mathbf{K}_j(B)=\mathbf{f}$ for every $j>i$ or
$\mathbf{K}_k(A)=\mathbf{t}$ for the smallest $k>i$ with $\mathbf{K}_k(B)=\mathbf{t}$
$\Leftrightarrow \mathbf{K}_i(A\ \mathtt{atnext}\ B)=\mathbf{t}$. □

All of the formulas (T1)–(T33) are implications or even equivalences. In classical logic any valid implication $A \rightarrow B$ can also be formulated as a relation:

$$A \Vdash B.$$

As we have seen, this is no longer true in temporal logic. Here we have another relationship because of Theorem 2.2: any law of the form $\Box A \rightarrow B$ is equivalently expressed by:

$$A \Vdash B.$$

An example of this case is (T26). It can also be represented in the form:

$$A \rightarrow B \Vdash \Box A \rightarrow \Box B.$$

We give a list of some more such laws in a representation with $\Vdash$, again proving only some examples.

Monotonicity laws

(T34) $A \rightarrow B \Vdash \circ A \rightarrow \circ B$
(T35) $A \rightarrow B \Vdash \Diamond A \rightarrow \Diamond B$
(T36) $A \rightarrow B \Vdash A\ \mathtt{atnext}\ C \rightarrow B\ \mathtt{atnext}\ C$

These relations express the *monotonicity* of $\circ$, $\Diamond$ and **atnext** (in the first argument) with respect to $\rightarrow$. Note that (T26) could also be taken among these.

Proof of (T34)

$\mathbf{K}_i(A \rightarrow B)=\mathbf{t}$ for every i
$\Rightarrow \mathbf{K}_i(A)=\mathbf{f}$ or $\mathbf{K}_i(B)=\mathbf{t}$ for every i
$\Rightarrow \mathbf{K}_{i+1}(A)=\mathbf{f}$ or $\mathbf{K}_{i+1}(B)=\mathbf{t}$ for every i
$\Rightarrow \mathbf{K}_i(\circ A)=\mathbf{f}$ or $\mathbf{K}_i(\circ B)=\mathbf{t}$ for every i
$\Rightarrow \mathbf{K}_i(\circ A \rightarrow \circ B)=\mathbf{t}$ for every i. □

Frame laws

(T37) $A \Vdash \circ B \rightarrow \circ (A \wedge B)$
(T38) $A \Vdash \Box B \rightarrow \Box (A \wedge B)$
(T39) $A \Vdash \Diamond B \rightarrow \Diamond (A \wedge B)$
(T40) $A \Vdash B\ \mathtt{atnext}\ C \rightarrow (A \wedge B)\ \mathtt{atnext}\ (A \wedge C)$

These relations mean that some valid A may be "added" as a conjunct under any other temporal operator.

Proof of (T37)

$$\begin{aligned}
&\mathbf{K}_i(A)=\mathbf{t} \quad \text{for every } i\\
&\quad\Rightarrow \mathbf{K}_{i+1}(A)=\mathbf{t} \quad \text{for every } i\\
&\quad\Rightarrow \mathbf{K}_{i+1}(B)=\mathbf{f} \quad \text{or}\\
&\qquad [\mathbf{K}_{i+1}(B)=\mathbf{t} \quad \text{and} \quad \mathbf{K}_{i+1}(A)=\mathbf{t}] \quad \text{for every } i\\
&\quad\Rightarrow \mathbf{K}_i(\circ B)=\mathbf{f} \quad \text{or} \quad \mathbf{K}_i(\circ(A \wedge B))=\mathbf{t} \quad \text{for every } i\\
&\quad\Rightarrow \mathbf{K}_i(\circ B \rightarrow \circ(A \wedge B))=\mathbf{t} \quad \text{for every } i. \quad \square
\end{aligned}$$

Temporal generalization laws

(T41) $\quad \Box A \rightarrow B \Vdash \Box A \rightarrow \Box B$

(T42) $\quad A \rightarrow \Diamond B \Vdash \Diamond A \rightarrow \Diamond B$

(T41) corresponds to the generalization law of first-order logic:

$$A \rightarrow B \Vdash A \rightarrow \forall x B \qquad (x \text{ not free in } A)$$

if we remember that the operator $\Box$ informally means some "for every i". (T42) is the dual for $\Diamond$.

Proof of (T41)

$$\begin{aligned}
&\mathbf{K}_i(\Box A \rightarrow B)=\mathbf{t} \quad \text{for every } i \quad \text{and} \quad \mathbf{K}_j(\Box A)=\mathbf{t}\\
&\quad\Rightarrow [\mathbf{K}_i(\Box A)=\mathbf{f} \quad \text{or} \quad \mathbf{K}_i(B)=\mathbf{t}] \quad \text{for every } i \quad \text{and}\\
&\qquad \mathbf{K}_k(A)=\mathbf{t} \quad \text{for every } k \geq j\\
&\quad\Rightarrow [\mathbf{K}_i(\Box A)=\mathbf{f} \quad \text{or} \quad \mathbf{K}_i(B)=\mathbf{t}] \quad \text{for every } i \geq j \quad \text{and}\\
&\qquad \mathbf{K}_k(\Box A)=\mathbf{t} \quad \text{for every } k \geq j\\
&\quad\Rightarrow \mathbf{K}_i(B)=\mathbf{t} \quad \text{for every } i \geq j\\
&\quad\Rightarrow \mathbf{K}_j(\Box B)=\mathbf{t}. \quad \square
\end{aligned}$$

4. Some Further Temporal Operators

We have already mentioned in Section 1 another reasonable temporal operator which has not yet occurred in the subsequent formal considerations:

A `until` B: "A holds at all following time points up to an (existing) time point at which B holds".

(Notice that we have now made precise that the time point at which B becomes true really exists.)

Actually there are also further reasonable operators which might be useful for particular applications. We want briefly to consider some of them:

A **unless** B: "If there is a following time point at which B holds then A holds up to that point or else A holds permanently" (*unless* or *weak until* operator),

A **while** B: "A holds as long as B holds (in the future)" (*while* operator),

A **before** B: "If B holds sometime in the future then A holds before that" (*precedence* or *before* operator).

The precise formal meaning of these operators is given by the definition of the semantic function $\mathbf{K}_i$ for such formulas:

$$\begin{array}{lll}
\mathbf{K}_i(A\ \mathtt{until}\ B)=\mathbf{t} & \text{iff} & \mathbf{K}_j(B)=\mathbf{t} \quad \text{for some } j>i \quad \text{and} \\
 & & \mathbf{K}_k(A)=\mathbf{t} \quad \text{for every } k,\ i<k<j. \\
\mathbf{K}_i(A\ \mathtt{unless}\ B)=\mathbf{t} & \text{iff} & [\mathbf{K}_j(B)=\mathbf{t} \quad \text{for some } j>i \quad \text{and} \\
 & & \mathbf{K}_k(A)=\mathbf{t} \quad \text{for every } k,\ i<k<j] \quad \text{or} \\
 & & \mathbf{K}_k(A)=\mathbf{t} \quad \text{for every } k>i. \\
\mathbf{K}_i(A\ \mathtt{while}\ B)=\mathbf{t} & \text{iff} & [\mathbf{K}_j(B)=\mathbf{f} \quad \text{for some } j>i \quad \text{and} \\
 & & \mathbf{K}_k(A)=\mathbf{K}_k(B)=\mathbf{t} \quad \text{for every } k,\ i<k<j] \quad \text{or} \\
 & & \mathbf{K}_k(A)=\mathbf{t} \quad \text{for every } k>i. \\
\mathbf{K}_i(A\ \mathtt{before}\ B)=\mathbf{t} & \text{iff} & \text{for every } j>i \text{ with } \mathbf{K}_j(B)=\mathbf{t} \\
 & & \text{there is some } k,\ i<k<j, \quad \text{with} \quad \mathbf{K}_k(A)=\mathbf{t}.
\end{array}$$

We have already seen that the operators $\circ$, $\Box$, $\Diamond$ can be expressed by the atnext operator. The same holds for these new operators (and, in fact, for a large class of other comparable operators as well). We list these equivalences as further logical laws:

(T43) $A\ \mathtt{until}\ B \leftrightarrow B\ \mathtt{atnext}\ (A \rightarrow B) \wedge \circ\Diamond B$

(T44) $A\ \mathtt{unless}\ B \leftrightarrow B\ \mathtt{atnext}\ (A \rightarrow B)$

(T45) $A\ \mathtt{while}\ B \leftrightarrow \neg B\ \mathtt{atnext}\ (A \rightarrow \neg B)$

(T46) $A\ \mathtt{before}\ B \leftrightarrow \neg B\ \mathtt{atnext}\ (A \vee B)$

Proof of (T44)

$\mathbf{K}_i(A\ \mathtt{unless}\ B)=\mathbf{t}$

$\Leftrightarrow$ [there is $j>i$ with $\mathbf{K}_j(B)=\mathbf{t}$ and $\mathbf{K}_k(A)=\mathbf{t}$ for $i<k<j$] or $\mathbf{K}_k(A)=\mathbf{t}$ for every $k>i$

$\Leftrightarrow$ [there is a smallest $j>i$ with $\mathbf{K}_j(B)=\mathbf{t}$ and $\mathbf{K}_k(A)=\mathbf{t}$ for $i<k<j$] or [$\mathbf{K}_k(A)=\mathbf{t}$ and $\mathbf{K}_k(B)=\mathbf{f}$] for every $k>i$

$\Leftrightarrow \mathbf{K}_k(A \rightarrow B)=\mathbf{f}$ for every $k>i$ or $\mathbf{K}_j(B)=\mathbf{t}$ for the smallest $j>i$ with $\mathbf{K}_j(A \rightarrow B)=\mathbf{t}$

$\Leftrightarrow \mathbf{K}_i(B\ \mathtt{atnext}\ (A \rightarrow B))=\mathbf{t}$. □

The proofs of the other equivalences are left to the reader. They show that the new operators are no proper extensions of the language $\mathscr{L}_{\mathrm{TA}}$. (T43)–(T46) may be viewed as their "definitions". The expressive power of **atnext** indicated by this fact can, however, also be found in each of these new operators itself. This means that each one could serve as a linguistic basis for $\mathscr{L}_{\mathrm{TA}}$ which may be seen by expressing **atnext** in terms of each of them:

$$A \,\mathtt{atnext}\, B \leftrightarrow \neg B \,\mathtt{until}\, (A \wedge B) \vee \circ\Box\neg B$$
$$A \,\mathtt{atnext}\, B \leftrightarrow \neg B \,\mathtt{unless}\, (A \wedge B)$$
$$A \,\mathtt{atnext}\, B \leftrightarrow \neg B \,\mathtt{while}\, (A \rightarrow \neg B)$$
$$A \,\mathtt{atnext}\, B \leftrightarrow B \,\mathtt{before}\, (\neg A \wedge B).$$

Proofs are again left to the reader.

We finally notice that the operators **atnext**, **unless**, **while** and **before** have a common property: they have no "existential" aspect, which means that, e.g., in A **unless** B, there need not exist a time point at which B becomes true. We call them *weak* operators. **until**, on the contrary, is a *strong* operator also expressing the existence of some time point. Such strong operators are not too interesting since they can always be expressed by their weak version and some $\Diamond$-formula expressing the existential part. For example:

$$A \,\mathtt{until}\, B \leftrightarrow A \,\mathtt{unless}\, B \wedge \circ\Diamond B.$$

Therefore, in subsequent developments of these operators we will concentrate on the weak ones and no longer follow up **until**.

Chapter II

Axiomatization of Propositional Temporal Logic

5. The Formal System Σ_{TA}

After the semantical definition of valid formulas of propositional temporal logic in Section 2, we now want to give a formal system Σ_{TA} for the formal derivation of such formulas:

Axioms

(taut)	All tautologically valid formulas,
(ax 1)	$\neg\circ A \leftrightarrow \circ\neg A$,
(ax 2)	$\circ(A \rightarrow B) \rightarrow (\circ A \rightarrow \circ B)$,
(ax 3)	$\Box A \rightarrow A \wedge \circ\Box A$,
(ax 4)	$\circ\Box\neg B \rightarrow A$ **atnext** B,
(ax 5)	A **atnext** $B \leftrightarrow \circ(B \rightarrow A) \wedge \circ(\neg B \rightarrow A$ **atnext** $B)$.

Rules

(mp)	$A, A \rightarrow B \vdash B$,
(nex)	$A \vdash \circ A$,
(ind)	$A \rightarrow B, A \rightarrow \circ A \vdash A \rightarrow \Box B$.

The "axiom" (taut) seems somewhat strange. We could also have taken some axioms for classical propositional logic. But since we are not interested here in how classical tautologies can be derived from such axioms we simply take all of them as axioms. In (ax 2) and (ax 3) one should notice that these are only implications and not equivalences although the equivalences are also valid according to (T 18) and (T 31). The rule (ind) is an *induction rule* with the informal meaning:

> "If A (always) implies B and A is invariant from any time point to the next then A implies B forever".

Let us now convince ourselves that Σ_{TA} allows for deriving only formulas which are valid.

Theorem 5.1. (Soundness Theorem for Σ_{TA}.) *Let A be a formula and $\mathcal{F}$ a set of formulas. If $\mathcal{F} \vdash A$ then $\mathcal{F} \Vdash A$. (In particular: if $\vdash A$ then $\Vdash A$.)*

Proof. The proof runs by induction on the assumed derivation of A from $\mathcal{F}$.

1. A is an axiom of Σ_{TA}. All axioms (taut), (ax 1), (ax 2), (ax 3), (ax 5) are valid as we have already seen in Theorem 3.1 and the laws (T 1), (T 18), (T 31) and (T 33).

An axiom (ax4) is also valid because

$$\begin{aligned} \mathbf{K}_i(\circ\Box\neg B)=\mathbf{t} &\Rightarrow \mathbf{K}_j(\neg B)=\mathbf{t} \quad \text{for every } j>i \\ &\Rightarrow \mathbf{K}_j(B)=\mathbf{f} \quad \text{for every } j>i \\ &\Rightarrow \mathbf{K}_i(A \;\mathtt{atnext}\; B)=\mathbf{t}. \end{aligned}$$

Hence we get in every case $\mathscr{F} \Vdash A$.

2. $A \in \mathscr{F}$. In this case we have trivially $\mathscr{F} \Vdash A$.
3. A is a conclusion of (mp) with premises B and $B \to A$. Then we have $\mathscr{F} \vdash B$ and $\mathscr{F} \vdash B \to A$, we get $\mathscr{F} \Vdash B$ and $\mathscr{F} \Vdash B \to A$ by the induction hypothesis, and hence $\mathscr{F} \Vdash A$ by Theorem 2.3.
4. A is a conclusion of (nex) with premise B. Then we have $\mathscr{F} \vdash B$ and $A \equiv \circ B$, we have $\mathscr{F} \Vdash B$ by the induction hypothesis, and $\mathscr{F} \Vdash \circ B$ by Theorem 2.4.
5. A is a conclusion of (ind) with premises $B \to C$ and $B \to \circ B$. Then we have $\mathscr{F} \vdash B \to C$, $\mathscr{F} \vdash B \to \circ B$ and $A \equiv B \to \Box C$, hence $\mathscr{F} \Vdash B \to C$ and $\mathscr{F} \Vdash B \to \circ B$ by the induction hypothesis. Let $\mathbf{K}$ be a temporal structure with $\Vdash_{\mathbf{K}} F$ for every $F \in \mathscr{F}$. Then

   i) $\Vdash_{\mathbf{K}} B \to C$ and
   ii) $\Vdash_{\mathbf{K}} B \to \circ B$.

   Let $i \in \mathbb{N}_0$ and $\mathbf{K}_i(B)=\mathbf{t}$. From ii) we get

   $$\mathbf{K}_{i+1}(B)=\mathbf{K}_{i+2}(B)=\mathbf{K}_{i+3}(B)=\ldots=\mathbf{t},$$

   i.e., $\mathbf{K}_j(B)=\mathbf{t}$ for every $j \geq i$. With i) we also get $\mathbf{K}_j(C)=\mathbf{t}$ for every $j \geq i$, hence together $\mathbf{K}_i(B \to \Box C)=\mathbf{t}$. This means $\mathscr{F} \Vdash A$. □

We argued above that in derivations within Σ_{TA} we do not want to be bothered about how to derive classical tautologies; we will simply use them as axioms. Similarly we would like to abbreviate purely classical derivation steps. Formally, we can perform this by using a derived rule:

(prop) $A_1, \ldots, A_n \vdash B$ whenever B is a tautological consequence of $A_1, \ldots, A_n$.

Again, we will use this rule very extensively without really proving the particular presupposition. The rule is justified by the following.

Theorem 5.2. *If B is a tautological consequence of $A_1, \ldots, A_n$ then $A_1, \ldots, A_n \vdash B$.*

Proof. We prove only the case $n=2$. The general case is analogous. If B is a tautological consequence of A_1 and A_2 then the formula $A_1 \wedge A_2 \to B$ is tautologically valid and we can give the following derivation of B from A_1 and A_2:

(1)	A_1	assumption
(2)	A_2	assumption
(3)	$A_1 \wedge A_2 \to B$	(taut)
(4)	$(A_1 \wedge A_2 \to B) \to (A_1 \to (A_2 \to B))$	(taut)
(5)	$A_1 \to (A_2 \to B)$	(mp), (3), (4)
(6)	$A_2 \to B$	(mp), (1), (5)
(7)	B	(mp), (2), (6) □

Observe in this proof our standard notation of a formal derivation. In each step (line) we list some derivable formula and indicate on the right-hand side by what axiom or rule and by what previous lines it is found.

Example. A very frequently occurring case of the rule (prop) is the *chain reasoning rule*:

$$A \rightarrow B, B \rightarrow C \vdash A \rightarrow C.$$

Of course, it denotes a tautological consequence and is therefore an instance of (prop). □

We now want to give an example of a derivation of a proper temporal formula:

(T18′) $(\circ A \rightarrow \circ B) \rightarrow \circ(A \rightarrow B)$.

(We have designated this formula by (T18′) because it is one part of (T18) – incidentally, just the part missing in (ax2).)

Derivation of (T18′)

(1)	$\neg(A \rightarrow B) \rightarrow A$	(taut)
(2)	$\circ(\neg(A \rightarrow B) \rightarrow A)$	(nex), (1)
(3)	$\circ(\neg(A \rightarrow B) \rightarrow A) \rightarrow (\circ\neg(A \rightarrow B) \rightarrow \circ A)$	(ax2)
(4)	$\circ\neg(A \rightarrow B) \rightarrow \circ A$	(mp), (2), (3)
(5)	$\neg\circ(A \rightarrow B) \leftrightarrow \circ\neg(A \rightarrow B)$	(ax1)
(6)	$\neg\circ(A \rightarrow B) \rightarrow \circ\neg(A \rightarrow B)$	(prop), (5)
(7)	$\neg\circ(A \rightarrow B) \rightarrow \circ A$	(prop), (4), (6)
(8)	$\neg(A \rightarrow B) \rightarrow \neg B$	(taut)
(9)	$\neg\circ(A \rightarrow B) \rightarrow \circ\neg B$	from (8) in the same way as (7) from (1)
(10)	$\circ\neg B \rightarrow \neg\circ B$	(prop), (ax1)
(11)	$\neg\circ(A \rightarrow B) \rightarrow \neg\circ B$	(prop), (9), (10)
(12)	$\neg\circ(A \rightarrow B) \rightarrow \neg(\circ A \rightarrow \circ B)$	(prop), (7), (11)
(13)	$(\circ A \rightarrow \circ B) \rightarrow \circ(A \rightarrow B)$	(prop), (12) □

Besides the rule (prop) we want to note another very useful derived rule:

(alw) $A \vdash \Box A$.

Derivation of (alw)

(1)	A	assumption
(2)	$\circ A$	(nex), (1)
(3)	$A \rightarrow \circ A$	(prop), (2)
(4)	$A \rightarrow A$	(taut)
(5)	$A \rightarrow \Box A$	(ind), (3), (4)
(6)	$\Box A$	(mp), (1), (5) □

In semantics we observed a connection between implication and the relation $\Vdash$. There is an analogous relationship between implication and derivability.

Theorem 5.3. (Deduction Theorem of propositional temporal logic.) *Let A, B be formulas, $\mathscr{F}$ a set of formulas. If $\mathscr{F} \cup \{A\} \vdash B$ then $\mathscr{F} \vdash \Box A \rightarrow B$.*

Proof. The proof runs by induction on the assumed derivation of B from $\mathscr{F} \cup \{A\}$.

1. B is an axiom of Σ_{TA} or $B \in \mathscr{F}$. Then we have $\mathscr{F} \vdash B$ and $\mathscr{F} \vdash B \rightarrow (\Box A \rightarrow B)$ by (taut) and hence $\mathscr{F} \vdash \Box A \rightarrow B$ by (mp).
2. $B \equiv A$. Then $\mathscr{F} \vdash \Box A \rightarrow A \wedge \circ \Box A$ by (ax 3), hence $\mathscr{F} \vdash \Box A \rightarrow A$ by (prop).
3. B is a conclusion of (mp) with premises C and $C \rightarrow B$. Then $\mathscr{F} \cup \{A\} \vdash C$ and $\mathscr{F} \cup \{A\} \vdash C \rightarrow B$, and $\mathscr{F} \vdash \Box A \rightarrow C$ and $\mathscr{F} \vdash \Box A \rightarrow (C \rightarrow B)$ by the induction hypothesis, and hence $\mathscr{F} \vdash \Box A \rightarrow B$ by (prop).
4. $B \equiv \circ C$ is a conclusion of (nex) with premise C. Then $\mathscr{F} \cup \{A\} \vdash C$ and by the induction hypothesis $\mathscr{F} \vdash \Box A \rightarrow C$. We give a derivation of $\Box A \rightarrow \circ C$ from $\Box A \rightarrow C$:

(1)	$\Box A \rightarrow C$	assumption
(2)	$\circ(\Box A \rightarrow C)$	(nex), (1)
(3)	$\circ(\Box A \rightarrow C) \rightarrow (\circ \Box A \rightarrow \circ C)$	(ax 2)
(4)	$\circ \Box A \rightarrow \circ C$	(mp), (2), (3)
(5)	$\Box A \rightarrow A \wedge \circ \Box A$	(ax 3)
(6)	$\Box A \rightarrow \circ \Box A$	(prop), (5)
(7)	$\Box A \rightarrow \circ C$	(prop), (4), (6)

5. $B \equiv C \rightarrow \Box D$ is a conclusion of (ind) with premises $C \rightarrow D$ and $C \rightarrow \circ C$. As above we get $\mathscr{F} \vdash \Box A \rightarrow (C \rightarrow D)$ and $\mathscr{F} \vdash \Box A \rightarrow (C \rightarrow \circ C)$ by the induction hypothesis. $C \rightarrow \Box D$ can be derived from these as follows:

(1)	$\Box A \rightarrow (C \rightarrow D)$	assumption
(2)	$\Box A \rightarrow (C \rightarrow \circ C)$	assumption
(3)	$\Box A \wedge C \rightarrow D$	(prop), (1)
(4)	$\Box A \wedge C \rightarrow \circ C$	(prop), (2)
(5)	$\Box A \rightarrow \circ \Box A$	(prop), (ax 3)
(6)	$\Box A \wedge C \rightarrow \circ \Box A \wedge \circ C$	(prop), (4), (5)
(7)	$\circ(\Box A \rightarrow \neg C) \rightarrow (\circ \Box A \rightarrow \circ \neg C)$	(ax 2)
(8)	$\circ(\Box A \rightarrow \neg C) \rightarrow (\circ \Box A \rightarrow \neg \circ C)$	(prop), (ax 1), (7)
(9)	$\circ \Box A \wedge \circ C \rightarrow \circ(\Box A \wedge C)$	(prop), (ax 1), (8)
(10)	$\Box A \wedge C \rightarrow \circ(\Box A \wedge C)$	(prop), (6), (9)
(11)	$\Box A \wedge C \rightarrow \Box D$	(ind), (3), (10)
(12)	$\Box A \rightarrow (C \rightarrow \Box D)$	(prop), (11) □

This theorem is formulated quite generally. Some special cases are the following:

i) If $A \vdash B$ then $\vdash \Box A \rightarrow B$.
ii) If $A_1, \ldots, A_n \vdash B$ then $\vdash \Box A_1 \wedge \ldots \wedge \Box A_n \rightarrow B$.

Observe again that the Deduction Theorem of classical propositional logic:

$$\text{If} \quad \mathscr{F} \cup \{A\} \vdash B \quad \text{then} \quad \mathscr{F} \vdash A \rightarrow B$$

does not hold here in general. However, as in classical logic, the Deduction Theorem can be used to abbreviate derivations of formulas. Let us illustrate this by a simple example:

(T26) $\quad \Box(A \to B) \to (\Box A \to \Box B)$

Derivation of (T 26)

By the Deduction Theorem it suffices to derive $A \to B \vdash \Box A \to \Box B$. For this, again by the Deduction Theorem, it suffices to derive $A \to B, A \vdash \Box B$. This is almost trivial.

(1)	$A \to B$	assumption
(2)	A	assumption
(3)	B	(mp), (1), (2)
(4)	$\Box B$	(alw), (3) □

We have formulated the Deduction Theorem only in the form "if ... then ..." and not as "... if and only if ...". In fact the opposite direction holds quite trivially:

Theorem 5.4. *Let A, B be formulas, $\mathscr{F}$ a set of formulas. If $\mathscr{F} \vdash \Box A \to B$ then $\mathscr{F} \cup \{A\} \vdash B$.*

Proof. If $\mathscr{F} \vdash \Box A \to B$ then also $\mathscr{F} \cup \{A\} \vdash \Box A \to B$ and $\mathscr{F} \cup \{A\} \vdash A$, hence $\mathscr{F} \cup \{A\} \vdash \Box A$ by (alw) and $\mathscr{F} \cup \{A\} \vdash B$ by (mp). □

We should observe, however, that this inversion of the Deduction Theorem also holds in the classical form:

If $\quad \mathscr{F} \vdash A \to B \quad$ then $\quad \mathscr{F} \cup \{A\} \vdash B$,

because this is nothing but an application of (mp).

We conclude this section by deriving four formulas which we will need in the following section:

(T 19′)	$\circ A \wedge \circ B \to \circ(A \wedge B)$
(ax 3′)	$A \wedge \circ\Box A \to \Box A$
(ax 4′)	$\neg(A\ \mathbf{atnext}\ B) \to \circ\Diamond B$
(ax 5′)	$\neg(A\ \mathbf{atnext}\ B) \to \circ(B \to \neg A) \wedge \circ(\neg B \to \neg(A\ \mathbf{atnext}\ B))$

Observe that (ax 5′) is one part of a "recursive equivalence" for $\neg(A\ \mathbf{atnext}\ B)$ in the same way as (ax 5) is for $A\ \mathbf{atnext}\ B$.

Derivation of (T 19′)

(1)	$\circ(A \to \neg B) \to (\circ A \to \circ\neg B)$	(ax 2)
(2)	$\circ\neg B \leftrightarrow \neg\circ B$	(ax 1)
(3)	$\circ(A \to \neg B) \to (\circ A \to \neg\circ B)$	(prop), (1), (2)
(4)	$\neg(\circ A \to \neg\circ B) \to \neg\circ(A \to \neg B)$	(prop), (3)
(5)	$\neg(\circ A \to \neg\circ B) \to \circ\neg(A \to \neg B)$	(prop), (ax 1), (4)
(6)	$\circ A \wedge \circ B \to \circ(A \wedge B)$	definition of $\wedge$ □

Derivation of (ax 3′)

(1)	$A \land \circ\Box A \rightarrow A$	(taut)
(2)	$\Box A \rightarrow A \land \circ\Box A$	(ax 3)
(3)	$\circ\Box A \rightarrow \circ(A \land \circ\Box A)$	(nex), (ax 2), (mp), (2)
(4)	$A \land \circ\Box A \rightarrow \circ(A \land \circ\Box A)$	(prop), (3)
(5)	$A \land \circ\Box A \rightarrow \Box A$	(ind), (1), (4) □

Derivation of (ax 4′)

(1)	$\circ\Box\neg B \rightarrow A \ \mathtt{atnext}\ B$	(ax 4)
(2)	$\neg(A \ \mathtt{atnext}\ B) \rightarrow \neg\circ\Box\neg B$	(prop), (1)
(3)	$\neg(A \ \mathtt{atnext}\ B) \rightarrow \circ\neg\Box\neg B$	(prop), (ax 1), (2)
(4)	$\neg(A \ \mathtt{atnext}\ B) \rightarrow \circ\Diamond B$	definition of $\Diamond$ □

Derivation of (ax 5′)

(1)	$A \ \mathtt{atnext}\ B \leftrightarrow \circ(B \rightarrow A) \land$ $\circ(\neg B \rightarrow A \ \mathtt{atnext}\ B)$	(ax 5)
(2)	$\neg(A \ \mathtt{atnext}\ B) \rightarrow \neg\circ(B \rightarrow A) \lor$ $\neg\circ(\neg B \rightarrow A \ \mathtt{atnext}\ B)$	(prop), (1)
(3)	$\neg(A \ \mathtt{atnext}\ B) \rightarrow \neg(\circ B \rightarrow \circ A) \lor$ $\neg(\circ\neg B \rightarrow \circ(A \ \mathtt{atnext}\ B))$	(prop), (T 18′), (2)
(4)	$\neg(A \ \mathtt{atnext}\ B) \rightarrow (\circ B \rightarrow \circ\neg A) \land$ $(\circ\neg B \rightarrow \circ\neg(A \ \mathtt{atnext}\ B))$	(prop), (ax 1), (3)
(5)	$\neg(A \ \mathtt{atnext}\ B) \rightarrow \circ(B \rightarrow \neg A) \land$ $\circ(\neg B \rightarrow \neg(A \ \mathtt{atnext}\ B))$	(prop), (T 18′), (4) □

6. Completeness of Σ_{TA}

We know the soundness of our formal system Σ_{TA}, i.e., that every derivable formula is valid. We now want to see that every valid formula is derivable. Our proof of this completeness of Σ_{TA} runs – in principle – along the proof idea in the classical case, often called the *Henkin-Hasenjäger method.* This general idea, however, has to be modified in many details for our situation. The proof consists in constructing appropriate sets of formulas and, throughout the proof, one main difference to the classical case will be that we have always to be very careful in order to get *finite* sets. In classical logic these sets may be infinite which makes the procedure much simpler.

We will note the steps in our proof by a series of lemmas. Let us first introduce the following notation. If $\mathscr{F}$ is a finite set of formulas we write:

$$\hat{\mathscr{F}} \text{ for } \begin{cases} A_1 \land \ldots \land A_n & \text{if } \mathscr{F} = \{A_1, \ldots, A_n\} \neq \emptyset, \\ \mathtt{true} & \text{if } \mathscr{F} = \emptyset. \end{cases}$$

Definition. A finite set $\mathscr{F}$ of formulas is called *inconsistent* if $\vdash \neg\hat{\mathscr{F}}$, otherwise it is called *consistent.*

Lemma 6.1. *Let $\mathscr{F}$ be a finite, consistent set of formulas and A some formula. Then $\mathscr{F} \cup \{A\}$ or $\mathscr{F} \cup \{\neg A\}$ is consistent.*

Proof. Assume both $\mathscr{F} \cup \{A\}$ and $\mathscr{F} \cup \{\neg A\}$ are inconsistent, i.e., $\vdash \neg(\hat{\mathscr{F}} \wedge A)$ and $\vdash \neg(\hat{\mathscr{F}} \wedge \neg A)$. By (prop) we get $\vdash \neg \hat{\mathscr{F}}$ which means that $\mathscr{F}$ is inconsistent and hence is a contradiction. □

Now let F be a formula. We define inductively a set $\tau(F)$ of formulas:

1. $\tau(v) = \{v\}$.
2. $\tau(\neg A) = \{\neg A\} \cup \tau(A)$.
3. $\tau(A \to B) = \{A \to B\} \cup \tau(A) \cup \tau(B)$.
4. $\tau(\circ A) = \{\circ A\}$.
5. $\tau(\Box A) = \{\Box A\} \cup \tau(A)$.
6. $\tau(A \text{ atnext } B) = \{A \text{ atnext } B\}$.

Informally, $\tau(F)$ can be viewed as the set of "subformulas" of F where, however, formulas $\circ A$ and A **atnext** B are treated as "indivisible".

If $\mathscr{F}$ is a set of formulas, then we let $\tau(\mathscr{F}) = \{A \mid A \in \tau(F),\ F \in \mathscr{F}\}$. Obviously, if $\mathscr{F}$ is finite then $\tau(\mathscr{F})$ is finite. Furthermore τ is "transitive" in the following sense:

$$\text{If} \quad A \in \tau(\mathscr{F}) \quad \text{and} \quad B \in \tau(A) \quad \text{then} \quad B \in \tau(\mathscr{F}).$$

Definition. A set $\mathscr{F}$ of formulas is called *complete* if for every formula $A \in \tau(\mathscr{F})$ either $A \in \mathscr{F}$ or $\neg A \in \mathscr{F}$ (but not both).

Lemma 6.2. *Let $\mathscr{F}$ be a finite, consistent and complete set of formulas.*
a) If $\vdash A \to B$, $A \in \mathscr{F}$, $B \in \tau(\mathscr{F})$ then $B \in \mathscr{F}$.
b) If $A \to B \in \tau(\mathscr{F})$ then: $A \to B \in \mathscr{F}$ iff $A \notin \mathscr{F}$ or $B \in \mathscr{F}$.

Proof. a) From $\vdash A \to B$ we get $\vdash \neg(A \wedge \neg B)$ by (prop). Assuming $\neg B \in \mathscr{F}$ we would get $\vdash \neg \hat{\mathscr{F}}$ by (prop) which cannot hold. Hence $\neg B \notin \mathscr{F}$ and $B \in \mathscr{F}$.
b) Let $A \to B \in \mathscr{F}$. We have $\vdash \neg(A \wedge (A \to B) \wedge \neg B)$ by (taut) and assuming $A \in \mathscr{F}$ and $B \notin F$, i.e., $\neg B \in \mathscr{F}$, we would get $\vdash \neg \hat{\mathscr{F}}$ by (prop). Hence $A \notin \mathscr{F}$ or $B \in \mathscr{F}$. If, on the other hand, $A \to B \notin \mathscr{F}$ then $\neg(A \to B) \in \mathscr{F}$. Now $\vdash \neg(\neg A(\to B) \wedge \neg A)$ and $\vdash \neg(\neg(A \to B) \wedge B)$ by (taut) and therefore $\neg A \notin \mathscr{F}$, i.e., $A \in \mathscr{F}$, and $B \notin \mathscr{F}$ as above. □

Lemma 6.3. *Let $\mathscr{F}$ be a finite, consistent set of formulas. There is a finite, consistent and complete set $\mathscr{F}^*$ of formulas with $\mathscr{F} \subset \mathscr{F}^*$.*

Proof. $\mathscr{F}^*$ is constructed by successively adding A or $\neg A$ to $\mathscr{F}$ for every $A \in \tau(\mathscr{F})$ according to which of these extensions is consistent. By Lemma 6.1 this is always possible. □

Let us call a set $\mathscr{F}^*$ according to Lemma 6.3 a *completion* of $\mathscr{F}$. In general, there may exist different completions of $\mathscr{F}$, but obviously only finitely many.

Lemma 6.4. *Let* $\mathscr{F}_1^*, \ldots, \mathscr{F}_n^*$ *be all different completions of some* $\mathscr{F}$. *Then* $\vdash \widehat{\mathscr{F}} \to \mathscr{F}_1^* \vee \ldots \vee \mathscr{F}_n^*$.

Proof. Let $\mathscr{F}_1', \ldots, \mathscr{F}_m'$ be all different sets $\mathscr{F}'$ with the property that either $A \in \mathscr{F}'$ or $\neg A \in \mathscr{F}'$ for every $A \in \tau(\mathscr{F})$. We first claim that:

$$(*) \qquad \vdash \widehat{\mathscr{F}_1'} \vee \ldots \vee \widehat{\mathscr{F}_m'},$$

and prove this by induction on the number of formulas in $\tau(\mathscr{F})$. For $\tau(\mathscr{F}) = \emptyset$ there is nothing to show since then $\mathscr{F}_1' = \emptyset$. If $\tau(\mathscr{F}) = \{A_1\}$ then $\mathscr{F}_1' = \{A_1\}$ and $\mathscr{F}_2' = \{\neg A_1\}$ and $\vdash A_1 \vee \neg A_1$ by (prop). Let now $\tau(\mathscr{F}) = \{A_1, A_2, \ldots, A_k\}$ and $\mathscr{F}_1'', \ldots, \mathscr{F}_l''$ be the respective combinations of A or $\neg A$ for $A \in \{A_2, \ldots, A_k\}$. Then $m = 2 \times l$ and

$$\mathscr{F}_1' = \mathscr{F}_1'' \cup \{A_1\}, \ldots, \mathscr{F}_l' = \mathscr{F}_l'' \cup \{A_1\},$$
$$\mathscr{F}_{l+1}' = \mathscr{F}_1'' \cup \{\neg A_1\}, \ldots, \mathscr{F}_m' = \mathscr{F}_l'' \cup \{\neg A_1\}.$$

By the induction hypothesis we have $\vdash \widehat{\mathscr{F}_1''} \vee \ldots \vee \widehat{\mathscr{F}_l''}$ and from that we get $\vdash \widehat{\mathscr{F}_1'} \vee \ldots \vee \widehat{\mathscr{F}_m'}$ by (prop). Now the sets $\mathscr{F}_1^*, \ldots, \mathscr{F}_n^*$ are just those $\mathscr{F}_i'$ for which $\mathscr{F}_i' \cup \mathscr{F}$ is consistent. Suppose these are $\mathscr{F}_1', \ldots, \mathscr{F}_n'$, i.e., $\mathscr{F}_i' \cup \mathscr{F}$ is inconsistent for $i > n$. Hence $\vdash \neg \widehat{\mathscr{F}_i' \cup \mathscr{F}}$ for $i > n$ and with (*) and (prop) we can infer $\vdash \widehat{\mathscr{F}} \to \mathscr{F}_1' \vee \ldots \vee \mathscr{F}_n'$, i.e., $\vdash \widehat{\mathscr{F}} \to \mathscr{F}_1^* \vee \ldots \vee \mathscr{F}_n^*$, from this. □

The informal meaning of some completion $\mathscr{F}^*$ of a consistent set $\mathscr{F}$ is that it gathers information about which subformulas of the formulas of $\mathscr{F}$ should be true in some state in order to get all formulas of $\mathscr{F}$ true in that same state. Let us illustrate this idea with a little example. Suppose $\mathscr{F}$ to consist of the single formula:

$$A \equiv (v_1 \to v_2) \to \Box v_3.$$

One possible completion of $\mathscr{F}$ is:

$$\mathscr{F}_1^* = \{A, v_1 \to v_2, \Box v_3, \neg v_1, v_2, v_3\}.$$

If all the (proper) parts of A in $\mathscr{F}_1^*$ are true (which is, in fact, possible because of the consistency), then A is true too. However, some of this information focussed on one state may also have implications for other states. In our example, $\Box v_3$ can only be true in a state if (v_3 is true in this state which is already noted by v_3 being in $\mathscr{F}_1^*$ and) v_3 is also true in every future state or, equivalently, $\Box v_3$ is true in the next state. The "transfer" of such information from one state to the next is the purpose of our next construction.

Let $\mathscr{F}$ be a set of formulas. We define a set $\sigma(\mathscr{F})$ of formulas as follows:

$$\sigma_1(\mathscr{F}) = \{A \mid \circ A \in \mathscr{F}\},$$
$$\sigma_2(\mathscr{F}) = \{\neg A \mid \neg \circ A \in \mathscr{F}\},$$
$$\sigma_3(\mathscr{F}) = \{\Box A \mid \Box A \in \mathscr{F}\},$$
$$\sigma_4(\mathscr{F}) = \{\neg \Box A \mid \neg \Box A \in \mathscr{F} \text{ and } A \in \mathscr{F}\},$$
$$\sigma_5(\mathscr{F}) = \{B \to A, \neg B \to A \text{ atnext } B \mid A \text{ atnext } B \in \mathscr{F}\},$$
$$\sigma_6(\mathscr{F}) = \{\neg \Box \neg B, B \to \neg A, \neg B \to \neg (A \text{ atnext } B) \mid \neg (A \text{ atnext } B) \in \mathscr{F}\},$$
$$\sigma(\mathscr{F}) = \sigma_1(\mathscr{F}) \cup \sigma_2(\mathscr{F}) \cup \sigma_3(\mathscr{F}) \cup \sigma_4(\mathscr{F}) \cup \sigma_5(\mathscr{F}) \cup \sigma_6(\mathscr{F}).$$

It is clear that if $\mathscr{F}$ is finite then so is $\sigma(\mathscr{F})$. Furthermore we have the following:

Lemma 6.5. *Let $\mathscr{F}$ be a finite set of formulas.*
a) $\vdash \hat{\mathscr{F}} \to \circ\widehat{\sigma(\mathscr{F})}$.
b) If $\mathscr{F}$ is consistent then so is $\sigma(\mathscr{F})$.

Proof. a) We show the following assertion: if $C \in \sigma(\mathscr{F})$ then $\vdash \hat{\mathscr{F}} \to \circ C$. The claim a) then follows immediately by (prop) and (T19′), formally derived at the end of the previous section. Let $C \in \sigma(\mathscr{F})$. We distinguish the six cases according to the definition of σ:

$C \in \sigma_1(\mathscr{F}) \quad \Rightarrow \circ C \in \mathscr{F}$
$\qquad \Rightarrow \vdash \hat{\mathscr{F}} \to \circ C$ — by (prop).

$C \equiv \neg A \in \sigma_2(\mathscr{F}) \quad \Rightarrow \neg \circ A \in \mathscr{F}$
$\qquad \Rightarrow \vdash \hat{\mathscr{F}} \to \neg \circ A$ — by (prop)
$\qquad \Rightarrow \vdash \hat{\mathscr{F}} \to \circ \neg A$ — with (ax 1).

$C \equiv \Box A \in \sigma_3(\mathscr{F}) \quad \Rightarrow \Box A \in \mathscr{F}$
$\qquad \Rightarrow \vdash \hat{\mathscr{F}} \to \Box A$ — by (prop)
$\qquad \Rightarrow \vdash \hat{\mathscr{F}} \to \circ \Box A$ — with (ax 3).

$C \equiv \neg \Box A \in \sigma_4(\mathscr{F}) \Rightarrow \neg \Box A \in \mathscr{F}$ and $A \in \mathscr{F}$
$\qquad \Rightarrow \vdash \hat{\mathscr{F}} \to \neg \Box A \wedge A$ — by (prop)
$\qquad \Rightarrow \vdash \hat{\mathscr{F}} \to \neg \circ \Box A$ — with (ax 3′)
$\qquad \Rightarrow \vdash \hat{\mathscr{F}} \to \circ \neg \Box A$ — with (ax 1).

$C \in \sigma_5(\mathscr{F})$, $\quad C \equiv B \to A$ or $C \equiv \neg B \to A$ **atnext** B
$\qquad \Rightarrow A$ **atnext** $B \in \mathscr{F}$
$\qquad \Rightarrow \vdash \hat{\mathscr{F}} \to A$ **atnext** B — by (prop)
$\qquad \Rightarrow \vdash \hat{\mathscr{F}} \to \circ C$ — with (ax 5) in both cases.

$C \in \sigma_6(\mathscr{F})$, $\quad C \equiv \neg \Box \neg B$ or $C \equiv B \to \neg A$ or
$\qquad C \equiv \neg B \to \neg (A$ **atnext** $B)$
$\qquad \Rightarrow \neg (A$ **atnext** $B) \in \mathscr{F}$
$\qquad \Rightarrow \vdash \hat{\mathscr{F}} \to \neg (A$ **atnext** $B)$ — by (prop)
$\qquad \Rightarrow \vdash \hat{\mathscr{F}} \to \circ C$ — with (ax 4′) and (ax 5′) in every case.

b) If $\sigma(\mathscr{F})$ is inconsistent then $\vdash \neg \widehat{\sigma(\mathscr{F})}$ and hence $\vdash \circ \neg \widehat{\sigma(\mathscr{F})}$ with (nex). With (ax 1) we get $\vdash \neg \circ \widehat{\sigma(\mathscr{F})}$, by a) we have $\vdash \neg \circ \widehat{\sigma(\mathscr{F})} \to \neg \hat{\mathscr{F}}$, and so we have $\vdash \neg \hat{\mathscr{F}}$ by (mp) which means that $\hat{\mathscr{F}}$ is inconsistent. □

Now let $\mathscr{F}$ be a finite, consistent and complete set of formulas. We define a tree $\mathsf{T}(\mathscr{F})$, the nodes of which are sets of formulas.

- The root of $\mathsf{T}(\mathscr{F})$ is $\mathscr{F}$.
- If $\mathscr{N}$ is a node of $\mathsf{T}(\mathscr{F})$ then the sons of $\mathscr{N}$ are all different completions of $\sigma(\mathscr{N})$.

According to our remarks and results, every node of $\mathsf{T}(\mathscr{F})$ is a finite, consistent and complete set of formulas. If $\mathscr{N}$ is a node then the subtree of $\mathsf{T}(\mathscr{F})$ with root $\mathscr{N}$ is just $\mathsf{T}(\mathscr{N})$.

Lemma 6.6. *Let $\mathscr{F}$ be a finite, consistent and complete set of formulas.*
a) $\mathsf{T}(\mathscr{F})$ *has only finitely many different nodes* $\mathscr{N}_1, \ldots, \mathscr{N}_n$.
b) $\vdash \widehat{\mathscr{N}}_1 \vee \ldots \vee \widehat{\mathscr{N}}_n \rightarrow \circ(\widehat{\mathscr{N}}_1 \vee \ldots \vee \widehat{\mathscr{N}}_n)$.

Proof. a) Completing some finite set $\mathscr{N}$ one uses formulas from the finite set $\tau(\mathscr{N})$. So it is only the operation σ which "produces" new formulas. But σ either decreases the number of symbols $\circ$ or **atnext** which is possible only finitely often, or for a new formula A we have $B \in \tau(A)$ for the formula B from which A arises. This consideration shows that there are only finitely many different formulas occurring in the whole tree $\mathsf{T}(\mathscr{F})$. Hence there are only finitely many different nodes in $\mathsf{T}(\mathscr{F})$.
b) By Lemma 6.5 a) we have $\vdash \widehat{\mathscr{N}}_i \rightarrow \circ\widehat{\sigma(\mathscr{N}_i)}$ for every $i=1, \ldots, n$. Let $\mathscr{N}'_{i1}, \ldots, \mathscr{N}'_{im}$ be all different completions of $\sigma(\mathscr{N}_i)$. By Lemma 6.4 we have $\vdash \widehat{\sigma(\mathscr{N}_i)} \rightarrow \widehat{\mathscr{N}'_{i1}} \vee \ldots \vee \widehat{\mathscr{N}'_{im}}$. Since $\mathscr{N}'_{ij} \in \{\mathscr{N}_1, \ldots, \mathscr{N}_n\}$, we also have $\vdash \widehat{\mathscr{N}'_{ij}} \rightarrow \widehat{\mathscr{N}}_1 \vee \ldots \vee \widehat{\mathscr{N}}_n$ for $j=1, \ldots, m$. So we get $\vdash \widehat{\sigma(\mathscr{N}_i)} \rightarrow \widehat{\mathscr{N}}_1 \vee \ldots \vee \widehat{\mathscr{N}}_n$ with (prop) and $\vdash \circ\widehat{\sigma(\mathscr{N}_i)} \rightarrow \circ(\widehat{\mathscr{N}}_1 \vee \ldots \vee \widehat{\mathscr{N}}_n)$ with (nex) and (ax2) and hence $\vdash \widehat{\mathscr{N}}_i \rightarrow \circ(\widehat{\mathscr{N}}_1 \vee \ldots \vee \widehat{\mathscr{N}}_n)$ for $i=1, \ldots, n$, from which the claim b) follows immediately. □

A path in $\mathsf{T}(\mathscr{F})$ is a sequence

$$\mathscr{F}_0, \mathscr{F}_1, \mathscr{F}_2, \ldots$$

of nodes such that $\mathscr{F}_0 = \mathscr{F}$ and $\mathscr{F}_{i+1}$ is a son of $\mathscr{F}_i$ for every i. In the terms of the informal discussion above, a path is one possible sequence of sets carrying the information about which formulas should be true in the single states of the state sequence following some reference state in order to get the formulas of $\mathscr{F}$ true in this latter state. Our last problem is that not every path is appropriate for our ultimate goal. Let us distinguish the appropriate paths by calling a path *complete* if the following condition holds for every i:

$$\text{If} \quad \neg\Box A \in \mathscr{F}_i \quad \text{then} \quad \neg A \in \mathscr{F}_j \quad \text{for some } j \geq i.$$

Lemma 6.7. *Let $\mathscr{F}$ be a finite, consistent and complete set of formulas, $\mathscr{F}_0, \mathscr{F}_1, \mathscr{F}_2, \ldots$ $\mathscr{F}_2, \ldots$ a complete path in $\mathsf{T}(\mathscr{F})$, and $i \in \mathbb{N}_0$.*
a) If $\circ A \in \tau(\mathscr{F}_i)$ then: $\circ A \in \mathscr{F}_i$ iff $A \in \mathscr{F}_{i+1}$.
b) If $\Box A \in \tau(\mathscr{F}_i)$ then: $\Box A \in \mathscr{F}_i$ iff $A \in \mathscr{F}_j$ for every $j \geq i$.
*c) If A **atnext** $B \in \tau(\mathscr{F}_i)$ then:*

*A **atnext** $B \in \mathscr{F}_i$ iff $B \notin \mathscr{F}_j$ for every $j > i$ or*
$A \in \mathscr{F}_k$ for the smallest $k > i$ with $B \in \mathscr{F}_k$.

Proof. a) If $\circ A \in \mathscr{F}_i$ then $A \in \sigma(\mathscr{F}_i)$, hence $A \in \mathscr{F}_{i+1}$. If $\circ A \notin \mathscr{F}_i$ then $\neg\circ A \in \mathscr{F}_i$, hence $\neg A \in \sigma(\mathscr{F}_i)$ and therefore $\neg A \in \mathscr{F}_{i+1}$ and $A \notin \mathscr{F}_{i+1}$.
b) Let $\Box A \in \mathscr{F}_i$. Then $A \in \tau(\mathscr{F}_i)$ and because of $\vdash \Box A \rightarrow A$ which follows from (ax3) we get $A \in \mathscr{F}_i$ with Lemma 6.2a). Furthermore, $\Box A \in \sigma(\mathscr{F}_i)$, and hence $\Box A \in \mathscr{F}_{i+1}$. By induction we may conclude that $A \in \mathscr{F}_j$ for every $j \geq i$. If, on the other hand, $A \in \mathscr{F}_j$ for every $j \geq i$ then $\neg\Box A \notin \mathscr{F}_i$ by the definition of complete paths from which we get $\Box A \in \mathscr{F}_i$.

c) Let $A\,\mathtt{atnext}\,B\in\mathscr{F}_i$ and not $B\notin\mathscr{F}_j$ for every $j>i$. Then there is a smallest $k>i$ with $B\in\mathscr{F}_k$. According to the construction of $\mathscr{F}_{i+1}$ we have $B\rightarrow A\in\mathscr{F}_{i+1}$ and $\neg B\rightarrow A\,\mathtt{atnext}\,B\in\mathscr{F}_{i+1}$. If $i+1<k$ we get $\neg B\in\mathscr{F}_{i+1}$ because of $B\in\tau(\mathscr{F}_{i+1})$ and therefore $A\,\mathtt{atnext}\,B\in\mathscr{F}_{i+1}$ with Lemma 6.2b). By induction we may conclude that $A\,\mathtt{atnext}\,B\in\mathscr{F}_{k-1}$ and $B\rightarrow A\in\mathscr{F}_k$ and finally $A\in\mathscr{F}_k$ by again applying Lemma 6.2b). Let, on the other hand, $A\,\mathtt{atnext}\,B\notin\mathscr{F}_i$. Then $\neg(A\,\mathtt{atnext}\,B)\in\mathscr{F}_i$ and therefore $\neg\Box\neg B\in\mathscr{F}_{i+1}$, $B\rightarrow\neg A\in\mathscr{F}_{i+1}$, and $\neg B\rightarrow\neg(A\,\mathtt{atnext}\,B)\in\mathscr{F}_{i+1}$ according to the construction. Then $\Box\neg B\notin\mathscr{F}_{i+1}$ and by b) we get $\neg B\notin\mathscr{F}_k$ for some $k>i$. Choose k minimal. Then $\neg B\in\mathscr{F}_j$ for $i<j<k$ and by induction and with Lemma 6.2b) we find $B\rightarrow\neg A\in\mathscr{F}_k$. Then we have $A\in\tau(\mathscr{F}_k)$, $B\in\tau(\mathscr{F}_k)$, $B\in\mathscr{F}_k$, and $\neg A\in\mathscr{F}_k$, and hence $A\notin\mathscr{F}_k$. This means that it is not $A\in\mathscr{F}_k$ for the smallest $k>i$ with $B\in\mathscr{F}_k$ (which actually exists). □

The properties of complete paths stated in this lemma will help us to find our desired result. We only have to guarantee that such a path always exists.

Lemma 6.8. *Let $\mathscr{F}$ be a finite, consistent and complete set of formulas. There is a complete path in* $\mathsf{T}(\mathscr{F})$.

Proof. We first show the following claim: If $\mathscr{N}$ is a node of $\mathsf{T}(\mathscr{F})$ and $\neg\Box A\in\mathscr{N}$ then there is a node $\mathscr{N}'$ *of* $\mathsf{T}(\mathscr{N})$ with $\neg A\in\mathscr{N}'$.
Assume $\neg A\notin\mathscr{N}'$ for every $\mathscr{N}'\in\mathsf{T}(\mathscr{N})$. Then we first have $A\in\mathscr{N}$, hence $\neg\Box A$ is contained in every son of $\mathscr{N}$. By induction we find $A\in\mathscr{N}'$ and hence $\vdash\widehat{\mathscr{N}'}\rightarrow A$ for every $\mathscr{N}'\in\mathsf{T}(\mathscr{N})$. Let $\mathscr{N}_1',\ldots,\mathscr{N}_n'$ be all nodes of $\mathsf{T}(\mathscr{N})$. We get $\vdash\widehat{\mathscr{N}_1'}\vee\ldots\vee\widehat{\mathscr{N}_n'}\rightarrow A$. By Lemma 6.6b) we have $\vdash\widehat{\mathscr{N}_1'}\vee\ldots\vee\widehat{\mathscr{N}_n'}\rightarrow\circ(\widehat{\mathscr{N}_1'}\vee\ldots\vee\widehat{\mathscr{N}_n'})$, so by (ind) we get $\vdash\widehat{\mathscr{N}_1'}\vee\ldots\vee\widehat{\mathscr{N}_n'}\rightarrow\Box A$. Because of $\mathscr{N}\in\{\mathscr{N}_1',\ldots,\mathscr{N}_n'\}$ we also have $\vdash\widehat{\mathscr{N}}\rightarrow\widehat{\mathscr{N}_1'}\vee\ldots\vee\widehat{\mathscr{N}_n'}$ and we get $\vdash\widehat{\mathscr{N}}\rightarrow\Box A$ by (prop). Because of $\neg\Box A\in\mathscr{N}$ this implies $\vdash\neg\widehat{\mathscr{N}}$ by (prop) which means that $\mathscr{N}$ is inconsistent. This is a contradiction, thus the claim is proved.

Now we can construct a complete path in the following way: let $\mathfrak{P}$ be an arbitrary path in $\mathsf{T}(\mathscr{F})$. If no node of $\mathfrak{P}$ contains formulas of the kind $\neg\Box A$ then $\mathfrak{P}$ is obviously complete.

Otherwise, let $\mathscr{N}_0$ be the first node in $\mathfrak{P}$ containing such a formula. We construct inductively a sequence $\mathscr{N}_0,\mathscr{N}_1,\mathscr{N}_2,\ldots$ of nodes of $\mathsf{T}(\mathscr{F})$: if $\mathscr{N}_k$ contains no $\neg\Box A$ then the sequence ends with $\mathscr{N}_k$; otherwise, let $\neg\Box A'\in\mathscr{N}_k$ be such that, in the construction of $\mathscr{N}_0,\ldots,\mathscr{N}_k$, $\neg\Box A'$ has not been used (in this way) more times than the other formulas of this kind in $\mathscr{N}_k$. By the above, there is $\mathscr{N}_{k+1}\in\mathsf{T}(\mathscr{N}_k)$ with $\neg A'\in\mathscr{N}_{k+1}$. If $\mathscr{N}_0,\mathscr{N}_1,\ldots$ is infinite then we get a path by taking the initial part of $\mathfrak{P}$ from $\mathscr{F}$ to $\mathscr{N}_0$ and then proceeding via $\mathscr{N}_1,\mathscr{N}_2$, and so on. If $\mathscr{N}_0,\ldots,\mathscr{N}_k$ is finite then we take the path from $\mathscr{F}$ to $\mathscr{N}_k$ in the same way and iterate the whole construction by choosing an arbitrary path in $\mathsf{T}(\mathscr{N}_k)$.

Let $\mathscr{F}_0,\mathscr{F}_1,\ldots$ be the resulting path and assume $\neg\Box A\in\mathscr{F}_i$ and $\neg A\notin\mathscr{F}_j$ for every $j\geq i$. Then $A\in\mathscr{F}_i$ and by induction (as above) we get $\neg\Box A\in\mathscr{F}_j$ for every $j\geq i$. Therefore $\mathscr{N}_0,\mathscr{N}_1,\ldots$ is infinite and there is some l such that, for every $k\geq l$, $\mathscr{N}_k$ is some $\mathscr{F}_j$, $j\geq i$, and $\neg\Box A\in\mathscr{N}_k$. By Lemma 6.6a) there are only finitely many formulas of this kind in $\mathsf{T}(\mathscr{F})$, hence $\neg\Box A$ will be used in the above construction of some $\mathscr{N}_{k+1}$, $k\geq l$. So we get $\neg A\in\mathscr{N}_{k+1}$ and this is a contradiction. Therefore, $\mathscr{F}_0,\mathscr{F}_1,\ldots$ is complete. □

Now we have in fact all means whereby we may prove a theorem which is a rather trivial transcription of the desired completeness result:

Theorem 6.9. (Satisfiability Theorem for Σ_{TA}.) *Every finite, consistent set of formulas is satisfiable.*

Proof. Let $\mathscr{F}$ be finite and consistent, $\mathscr{F}^*$ some completion of $\mathscr{F}$, and $\mathscr{F}_0, \mathscr{F}_1, \mathscr{F}_2, \dots$ a complete path in $\mathsf{T}(\mathscr{F}^*)$. We define a temporal structure $\mathbf{K} = \{\eta_0, \eta_1, \eta_2, \dots\}$ by:

$$\eta_i(v) = \mathbf{t} \quad \text{iff} \quad v \in \mathscr{F}_i \qquad \text{for every } v \in \mathscr{V}.$$

We have to show that $\mathbf{K}_i(A) = \mathbf{t}$ for every $A \in \mathscr{F}$ and some $i \in \mathbb{N}_0$, and because of $\mathscr{F} \subset \mathscr{F}^* = \mathscr{F}_0$ this is certainly true for $i = 0$ if we have proved the following claim:

$$\text{If} \quad F \in \tau(\mathscr{F}_i) \quad \text{then:} \quad \mathbf{K}_i(F) = \mathbf{t} \quad \text{iff} \quad F \in \mathscr{F}_i.$$

The proof of this runs by induction on F:

1. $F \equiv v$: In this case the claim is just the definition of $\mathbf{K}$.
2. $F \equiv \neg A$: $\mathbf{K}_i(F) = \mathbf{t} \Leftrightarrow \mathbf{K}_i(A) = \mathbf{f}$
   $\Leftrightarrow A \notin \mathscr{F}_i$ by ind. hyp.
   $\Leftrightarrow F \in \mathscr{F}_i$.
3. $F \equiv A \to B$: $\mathbf{K}_i(F) = \mathbf{t} \Leftrightarrow \mathbf{K}_i(A) = \mathbf{f}$ or $\mathbf{K}_i(B) = \mathbf{t}$
   $\Leftrightarrow A \notin \mathscr{F}_i$ or $B \in \mathscr{F}_i$ by ind. hyp.
   $\Leftrightarrow F \in \mathscr{F}_i$ by Lemma 6.2b).
4. $F \equiv \circ A$: $\mathbf{K}_i(F) = \mathbf{t} \Leftrightarrow \mathbf{K}_{i+1}(A) = \mathbf{t}$
   $\Leftrightarrow A \in \mathscr{F}_{i+1}$ by ind. hyp.
   $\Leftrightarrow F \in \mathscr{F}_i$ by Lemma 6.7a).
5. $F \equiv \Box A$: $\mathbf{K}_i(F) = \mathbf{t} \Leftrightarrow \mathbf{K}_j(A) = \mathbf{t}$ for every $j \geq i$
   $\Leftrightarrow A \in \mathscr{F}_j$ for every $j \geq i$ by ind. hyp.
   $\Leftrightarrow F \in \mathscr{F}_i$ by Lemma 6.7b).
6. $F \equiv A$ B: $\mathbf{K}_i(F) = \mathbf{t}$
   $\Leftrightarrow \mathbf{K}_j(B) = \mathbf{f}$ for every $j > i$ or
   $\mathbf{K}_k(A) = \mathbf{t}$ for the smallest $k > i$ with $\mathbf{K}_k(B) = \mathbf{t}$
   $\Leftrightarrow B \notin \mathscr{F}_j$ for every $j > i$ or
   $A \in \mathscr{F}_k$ for the smallest $k > i$ with $B \in \mathscr{F}_k$ by ind. hyp.
   $\Leftrightarrow F \in \mathscr{F}_i$ by Lemma 6.7c). □

Theorem 6.10. (Completeness Theorem for Σ_{TA}.) *For every formula A, if $\Vdash A$ then $\vdash A$.*

Proof. If $\Vdash A$ then $\{\neg A\}$ is not satisfiable by Theorem 2.5 and inconsistent by Theorem 6.9. This means $\vdash \neg\neg A$ from which we get $\vdash A$ by (prop). □

Let us summarize. We now know from the Soundness and the Completeness Theorems that the valid and the derivable formulas are just the same:

$$\Vdash A \quad \text{iff} \quad \vdash A.$$

This means, in particular, that we can view all logical laws (T1)–(T33) as derivable and we will in fact use them in subsequent derivations.

We also know by the Soundness Theorem that:

$$\text{if} \quad \mathscr{F} \vdash A \quad \text{then} \quad \mathscr{F} \Vdash A$$

but it should be noticed that the converse relation

$$\text{if} \quad \mathscr{F} \Vdash A \quad \text{then} \quad \mathscr{F} \vdash A$$

does *not* hold in general. A simple counterexample is given by the infinite set $\mathscr{F} = \{A \rightarrow B,\ A \rightarrow \circ B,\ A \rightarrow \circ\circ B,\ A \rightarrow \circ\circ\circ B, \ldots\}$. It should be intuitively clear that $\mathscr{F} \Vdash A \rightarrow \Box B$, but $A \rightarrow \Box B$ cannot be derived from $\mathscr{F}$. (The reason is that in any derivation we could use only finitely many premises from $\mathscr{F}$.)

Here we have one more major difference to classical logic where this stronger completeness assertion holds. This difference is also expressed in our Satisfiability Theorem which holds only for finite sets of formulas but not for infinite ones as in classical logic. These remarks, however, suggest that at least the following holds in temporal logic:

$$\text{If} \quad \mathscr{F} \text{ finite} \quad \text{and} \quad \mathscr{F} \Vdash \mathrm{A} \quad \text{then} \quad \mathscr{F} \vdash A.$$

This, in fact, is proved very easily. If $\mathscr{F} = \{A_1, \ldots, A_n\}$ then

$$\begin{aligned} \{A_1, \ldots, A_n\} \Vdash A &\Rightarrow \Vdash \Box A_1 \wedge \ldots \wedge \Box A_n \rightarrow A && \text{by Theorem 2.2} \\ &\Rightarrow \vdash \Box A_1 \wedge \ldots \wedge \Box A_n \rightarrow A && \text{by Theorem 6.10} \\ &\Rightarrow A_1, \ldots, A_n \vdash A && \text{by Theorem 5.4.} \end{aligned}$$

This finally shows that we can use the logical laws (T34)–(T42) as derived rules, for example,

(T34) $\quad A \rightarrow B \vdash \circ A \rightarrow \circ B$,

because in every case the set of premises is finite.

We conclude this section by a first illustration of the use of the laws (T1)–(T42) in formal derivations. We prove the following additional distributivity law

(T47) $\quad \Box\Diamond(A \vee B) \leftrightarrow \Box\Diamond A \vee \Box\Diamond B$

which will be needed in later sections. One direction of (T47) is trivial:

(1)	$\Box\Diamond A \vee \Box\Diamond B \rightarrow \Box(\Diamond A \vee \Diamond B)$	(T27)
(2)	$\Diamond A \vee \Diamond B \rightarrow \Diamond(A \vee B)$	(prop), (T23)
(3)	$\Box\Diamond A \vee \Box\Diamond B \rightarrow \Box\Diamond(A \vee B)$	(prop), (T26), (1), (2)

For the other direction it suffices because of the Deduction Theorem to derive $\Diamond(A \vee B) \vdash \Box\Diamond A \vee \Box\Diamond B$:

(4)	$\Diamond(A \vee B)$	assumption
(5)	$\Diamond\Box\neg A \rightarrow \Diamond(\Diamond(A \vee B) \wedge \Box\neg A)$	(T39), (4)
(6)	$\Box\neg A \rightarrow (\Diamond(A \vee B) \rightarrow \Diamond(\neg A \wedge (A \vee B)))$	(T39)
(7)	$\Diamond(A \vee B) \wedge \Box\neg A \rightarrow \Diamond(\neg A \wedge (A \vee B))$	(prop), (6)

(8)	$\neg A \wedge (A \vee B) \rightarrow B$	(taut)
(9)	$\Diamond(\Diamond(A \vee B) \wedge \Box\neg A) \rightarrow \Diamond B$	(prop), (T35), (T42), (7), (8)
(10)	$\Diamond\Box\neg A \rightarrow \Diamond B$	(prop), (5), (9)
(11)	$\Box\Diamond\Box\neg A \rightarrow \Box\Diamond B$	(T26), (10)
(12)	$\Box\neg A \rightarrow \Box\Box\neg A$	(prop), (T14)
(13)	$\Diamond\Box\neg A \rightarrow \Diamond\Box\Box\neg A$	(T35), (12)
(14)	$\Diamond\Box\Box\neg A \rightarrow \Box\Diamond\Box\neg A$	(T10)
(15)	$\Diamond\Box\neg A \rightarrow \Box\Diamond B$	(prop), (11), (13), (14)
(16)	$\Box\Diamond A \vee \Box\Diamond B$	(prop), (T2), (T3), (15) □

7. Induction Principles

In the formal system Σ_{TA} every valid formula can be derived. We now want to investigate general proof *principles* or *strategies* for deriving some special kinds of formula. Let us begin with formulas of the kind

$$A \rightarrow \Box B.$$

The rule

(ind) $\quad A \rightarrow B,\ A \rightarrow \circ A \vdash A \rightarrow \Box B$

contained in Σ_{TA} is itself a general principle for proving $A \rightarrow \Box B$; more precisely it is an *induction principle* expressing that in order to be sure about $A \rightarrow \Box B$ one has to check that A implies B and is an invariant of any state transition. We first want to formulate two variants of (ind) which are more useful for subsequent applications:

(ind') $\quad A \rightarrow \circ A \vdash A \rightarrow \Box A,$

(ind'') $\quad A \rightarrow B,\ B \rightarrow \circ B \vdash A \rightarrow \Box B.$

Derivation of (ind')

(1)	$A \rightarrow \circ A$	assumption
(2)	$A \rightarrow A$	(taut)
(3)	$A \rightarrow \Box A$	(ind), (1), (2) □

Derivation of (ind'')

(1)	$A \rightarrow B$	assumption
(2)	$B \rightarrow \circ B$	assumption
(3)	$B \rightarrow \Box B$	(ind'), (2)
(4)	$A \rightarrow \Box B$	(prop), (1), (3) □

Let us consider next an even more specialized kind of formula, viz.,

$$A \rightarrow \Box(B \rightarrow C).$$

Informally this expresses "A implies that C will hold whenever B holds", and, of course, it can be proved by, say, (ind'') by proving that $B \rightarrow C$ is an invariant (and

A implies $B \to C$). However, it should be intuitively clear that this formula is also valid if we can guarantee that

i) "A implies C at the next time that B holds", and
ii) "if B and C hold then C will hold at the next time that B holds".

We can formally express this proof principle by:

(gind) $\quad A \to C \;\mathbf{atnext}\; B,\; B \wedge C \to C \;\mathbf{atnext}\; B \vdash A \to \Box\circ(B \to C)$

and call it the *generalized induction principle.* Notice that the conclusion is $A \to \Box\circ(B \to C)$ and not really $A \to \Box(B \to C)$ since the atnext operator does not include the present. In order to get the latter conclusion we would have to add the premise $A \to (B \to C)$.

Derivation of (gind)

(1)	$A \to C \;\mathbf{atnext}\; B$	assumption
(2)	$B \wedge C \to C \;\mathbf{atnext}\; B$	assumption
(3)	$C \;\mathbf{atnext}\; B \to \circ(B \to C)$ $\wedge \circ(\neg B \to C \;\mathbf{atnext}\; B)$	(prop), (ax 5)
(4)	$C \;\mathbf{atnext}\; B \wedge \circ B \to \circ(B \wedge C)$	(prop), (3)
(5)	$C \;\mathbf{atnext}\; B \wedge \circ B \to \circ(C \;\mathbf{atnext}\; B)$	(prop), (T 34), (2), (4)
(6)	$C \;\mathbf{atnext}\; B \wedge \neg\circ B \to \circ(C \;\mathbf{atnext}\; B)$	(prop), (3)
(7)	$C \;\mathbf{atnext}\; B \to \circ(C \;\mathbf{atnext}\; B)$	(prop), (5), (6)
(8)	$A \to \Box(C \;\mathbf{atnext}\; B)$	(ind''), (1), (7)
(9)	$A \to \Box\circ(B \to C)$	(prop), (T 26), (3), (8) □

We now turn to another class of formulas, those of the form:

$$A \to B \;\mathbf{atnext}\; C$$

In this case, Σ_{TA} does not contain much useful information about how to prove such a formula. Besides the trivial axiom $\circ\Box\neg C \to B \;\mathbf{atnext}\; C$ there is only the recursive characterization:

$$B \;\mathbf{atnext}\; C \leftrightarrow \circ(C \to B) \wedge \circ(\neg C \to B \;\mathbf{atnext}\; C)$$

which obviously does not help directly for proving the formula above. However, we can derive another induction principle from this recursion which is much better applicable:

(indatnext) $\quad A \to \circ(C \to B) \wedge \circ(\neg C \to A) \vdash A \to B \;\mathbf{atnext}\; C.$

Derivation

(1)	$A \to \circ(C \to B) \wedge \circ(\neg C \to A)$	assumption
(2)	$\neg(B \;\mathbf{atnext}\; C) \to$ $\circ(C \to \neg B) \wedge \circ(\neg C \to \neg(B \;\mathbf{atnext}\; C))$	(ax 5')
(3)	$A \wedge \neg(B \;\mathbf{atnext}\; C) \to \circ\neg C$	(prop), (1), (2)
(4)	$A \wedge \neg(B \;\mathbf{atnext}\; C) \to \circ(A \wedge \neg(B \;\mathbf{atnext}\; C))$	(prop), (1), (2), (3)
(5)	$A \wedge \neg(B \;\mathbf{atnext}\; C) \to \Box\circ\neg C$	(ind), (3), (4)
(6)	$\circ\Box\neg C \to B \;\mathbf{atnext}\; C$	(ax 4)

(7) $A \wedge \neg(B\,\mathbf{atnext}\,C) \rightarrow B\,\mathbf{atnext}\,C$ (prop), (T16), (5), (6)
(8) $A \rightarrow B\,\mathbf{atnext}\,C$ (prop), (7) □

We want briefly to illustrate the use of this induction principle by deriving a further useful rule:

(T48) $B \rightarrow A \vdash A\,\mathbf{atnext}\,B$

Derivation

(1)	$B \rightarrow A$	assumption
(2)	$\circ(B \rightarrow A)$	(nex), (1)
(3)	$\neg B \rightarrow (B \rightarrow A)$	(taut)
(4)	$\circ(\neg B \rightarrow (B \rightarrow A))$	(nex), (3)
(5)	$(B \rightarrow A) \rightarrow \circ(B \rightarrow A) \wedge \circ(\neg B \rightarrow (B \rightarrow A))$	(prop), (2), (4)
(6)	$(B \rightarrow A) \rightarrow A\,\mathbf{atnext}\,B$	(indatnext), (5)
(7)	$A\,\mathbf{atnext}\,B$	(mp), (1), (6) □

With this rule one can derive some simple **atnext** formulas, e.g.,

$$A\,\mathbf{atnext}\,A,$$
$$\Box A \rightarrow A\,\mathbf{atnext}\,B.$$

The first one follows immediately because of $\vdash A \rightarrow A$ (taut). Furthermore, we have $A \vdash (B \rightarrow A)$ and so we get $A \vdash A\,\mathbf{atnext}\,B$ with (T48) and hence the second formula with the Deduction Theorem.

The rule (indatnext) can also be very easily extended to the iterations of **atnext** for example,

(indatnext²) $A \rightarrow \circ(C \rightarrow B_1) \wedge \circ(\neg C \rightarrow A)$,
$B_1 \rightarrow \circ(C \rightarrow B) \wedge \circ(\neg C \rightarrow B_1)$
$\vdash A \rightarrow B\,\mathbf{atnext}^2\,C.$

Derivation

(1)	$A \rightarrow \circ(C \rightarrow B_1) \wedge \circ(\neg C \rightarrow A)$	assumption
(2)	$B_1 \rightarrow \circ(C \rightarrow B) \wedge \circ(\neg C \rightarrow B_1)$	assumption
(3)	$B_1 \rightarrow B\,\mathbf{atnext}\,C$	(indatnext), (2)
(4)	$A \rightarrow \circ(C \rightarrow B\,\mathbf{atnext}\,C) \wedge \circ(\neg C \rightarrow A)$	(prop), (T34), (1), (3)
(5)	$A \rightarrow (B\,\mathbf{atnext}\,C)\,\mathbf{atnext}\,C$	(indatnext), (4) □

This rule obviously extends to the general case:

(indatnextⁿ) $A \rightarrow \circ(C \rightarrow B_1) \wedge \circ(\neg C \rightarrow A)$,
$B_1 \rightarrow \circ(C \rightarrow B_2) \wedge \circ(\neg C \rightarrow B_1)$,
⋮
$B_{n-1} \rightarrow \circ(C \rightarrow B) \wedge \circ(\neg C \rightarrow B_{n-1})$
$\vdash A \rightarrow B\,\mathbf{atnext}^n\,C,$

which can be derived analogously.

Finally we can transfer the rule (indatnext) to the operators **unless**, **while** and **before** defined in terms of **atnext** in Section 4. We get further induction rules for these operators:

(indunless)	$A \rightarrow \circ C \vee \circ(A \wedge B)$	$\vdash A \rightarrow B$ **unless** C,
(indwhile)	$A \rightarrow \circ(C \rightarrow A \wedge B)$	$\vdash A \rightarrow B$ **while** C,
(indbefore)	$A \rightarrow \circ\neg C \wedge \circ(A \vee B)$	$\vdash A \rightarrow B$ **before** C.

We give only the derivation of (indunless).

Derivation of (indunless)

(1)	$A \rightarrow \circ C \vee \circ(A \wedge B)$	assumption
(2)	$C \vee (A \wedge B) \rightarrow ((B \rightarrow C) \rightarrow C) \wedge (\neg(B \rightarrow C) \rightarrow A)$	(taut)
(3)	$A \rightarrow \circ((B \rightarrow C) \rightarrow C) \wedge \circ(\neg(B \rightarrow C) \rightarrow A)$	(prop), (T19), (T20), (T34), (1), (2)
(4)	$A \rightarrow C$ **atnext** $(B \rightarrow C)$	(indatnext), (3)
(5)	$A \rightarrow B$ **unless** C	(T44), (4) □

The derivation of the other rules is analogous.

We have now investigated induction principles for formulas of the kind:

$A \rightarrow \Box B$
$A \rightarrow B$ **atnext** C

and – derived – $A \rightarrow B$ **op** C for the derived operators **op** $\equiv$ **unless**, **while** and **before**. We are only missing a last class of formulas of similar structure – formulas of the kind:

$A \rightarrow \Diamond B.$

However, there is no comparable proof principle for such formulas. Let us give at least a vague informal argument for this. The formula $A \rightarrow \Diamond B$ is equivalent to the formula $\neg\Diamond B \rightarrow \neg A$ and this is the same as $\Box\neg B \rightarrow \neg A$. Thus by the Deduction Theorem and its converse, the problem of deriving $A \rightarrow \Diamond B$ amounts to the problem of proving $\neg B \vdash \neg A$ or, say,

$B' \vdash A'.$

We cannot expect to be able to formulate a single proof principle as a rule within Σ_{TA} for deriving A' from some B'. Thus, although there must exist a derivation for $A \rightarrow \Diamond B$ if it is valid, we cannot expect to have a general device for this proof formulated in Σ_{TA}. If anything, we can hope to find some general method adding some means from outside the system. We will come back to this in Section 10.

Chapter III
First-Order Temporal Logic

8. First-Order Temporal Languages and Their Semantics

Temporal logic can be developed from its propositional basis to a first-order predicate logic in a way analogous to how this is done in the classical case.

Let $\mathscr{L}_P$ be a classical first-order language. A *first-order temporal language* $\mathscr{L}_{TP}$ (*with kernel* $\mathscr{L}_P$) is an extension of $\mathscr{L}_P$ as follows: the alphabet of $\mathscr{L}_{TP}$ is that of $\mathscr{L}_P$ with the additional symbols $\circ$, $\Box$ and **atnext**. Moreover, we assume the set of variables to be partitioned into two subsets which we call the set of *global* and *local* variables, respectively. *Terms* and *atomic formulas* are defined as in $\mathscr{L}_P$.

Inductive definition of *formulas*

1. Every atomic formula is a formula.
2. If A and B are formulas then $\neg A$, $(A \rightarrow B)$, $\circ A$, $\Box A$ and (A **atnext** B) are formulas.
3. If A is a formula and x is a global variable then $\forall x A$ is a formula.

We take over all conventions and definitions concerning further logical operators, parentheses, free and bound variables, substitution, etc., from previous sections. Furthermore, we call a formula of $\mathscr{L}_{TP}$ *closed* if it contains no free global variables. If $x_1, \ldots, x_n$ are all free global variables of some formula A then the formula $\forall x_1 \ldots \forall x_n A$ is called the *universal closure* of A.

The semantics of $\mathscr{L}_{TP}$ is given by extending the notion of temporal structure. A (*first-order*) *temporal structure* $\mathbf{K}=(\mathbf{S}, \xi, \mathbf{W})$ for $\mathscr{L}_{TP}$ consists of:

- a structure $\mathbf{S}$ for the kernel $\mathscr{L}_P$ of $\mathscr{L}_{TP}$,
- a global variable valuation ξ with respect to $\mathbf{S}$ (i.e., an assignment of an element of $|\mathbf{S}|$ to every global variable),
- an infinite sequence $\mathbf{W}=\{\eta_0, \eta_1, \eta_2, \ldots\}$ of *states* where each η_i is a local variable valuation with respect to $\mathbf{S}$ (i.e., an assignment of an element of $|\mathbf{S}|$ to every local variable).

In any $\mathbf{K}$, $\mathbf{S}$ and ξ together with $\eta_i \in \mathbf{W}$ define a value $\mathbf{S}^{(\xi,\eta_i)}(t) \in |\mathbf{S}|$ for every term t and a value $\mathbf{S}^{(\xi,\eta_i)}(A) \in \{\mathbf{f}, \mathbf{t}\}$ for every atomic formula exactly as in classical logic:

- $\mathbf{S}^{(\xi,\eta_i)}(x) = \xi(x)$ for every global variable x,
- $\mathbf{S}^{(\xi,\eta_i)}(a) = \eta_i(a)$ for every local variable a,
- $\mathbf{S}^{(\xi,\eta_i)}(f(t_1, \ldots, t_n)) = \mathbf{S}(f)(\mathbf{S}^{(\xi,\eta_i)}(t_1), \ldots, \mathbf{S}^{(\xi,\eta_i)}(t_n))$,
- $\mathbf{S}^{(\xi,\eta_i)}(p(t_1, \ldots, t_n)) = \mathbf{t}$ iff $(\mathbf{S}^{(\xi,\eta_i)}(t_1), \ldots, \mathbf{S}^{(\xi,\eta_i)}(t_n)) \in \mathbf{S}(p)$ for p other than $=$,
- $\mathbf{S}^{(\xi,\eta_i)}(t_1 = t_2) = \mathbf{t}$ iff $\mathbf{S}^{(\xi,\eta_i)}(t_1) \underset{|\mathbf{S}|}{=} \mathbf{S}^{(\xi,\eta_i)}(t_2)$.

$\mathbf{S}^{(\xi, \eta_i)}$ now plays the role of η_i in the case of propositional temporal logic and again we can inductively define $\mathbf{K}_i(F) \in \{\mathbf{f}, \mathbf{t}\}$ for every $\mathbf{K} = (\mathbf{S}, \xi, \mathbf{W})$, formula F and $i \in \mathbb{N}_0$:

1. $\mathbf{K}_i(A) = \mathbf{S}^{(\xi, \eta_i)}(A)$ for every atomic formula.
2. $\mathbf{K}_i(\neg A) = \mathbf{t}$ iff $\mathbf{K}_i(A) = \mathbf{f}$.
3. $\mathbf{K}_i(A \rightarrow B) = \mathbf{t}$ iff $\mathbf{K}_i(A) = \mathbf{f}$ or $\mathbf{K}_i(B) = \mathbf{t}$.
4. $\mathbf{K}_i(\circ A) = \mathbf{t}$ iff $\mathbf{K}_{i+1}(A) = \mathbf{t}$.
5. $\mathbf{K}_i(\Box A) = \mathbf{t}$ iff $\mathbf{K}_j(A) = \mathbf{t}$ for every $j \geq i$.
6. $\mathbf{K}_i(A \,\mathtt{atnext}\, B) = \mathbf{t}$ iff $\mathbf{K}_j(B) = \mathbf{f}$ for every $j > i$ or $\mathbf{K}_k(A) = \mathbf{t}$ for the smallest $k > i$ with $\mathbf{K}_k(B) = \mathbf{t}$.
7. $\mathbf{K}_i(\forall x A) = \mathbf{t}$ iff $\mathbf{K}'_i(A) = \mathbf{t}$ for every temporal structure $\mathbf{K}' = (\mathbf{S}, \xi', \mathbf{W})$ with $\xi'(y) \underset{|\mathbf{S}|}{=} \xi(y)$ for every y other than x.

Observe that lines 2–6 are the same as in the propositional case. Line 7 is the obvious extension of the respective classical definition. The definitions are transferred to the other propositional operators as before. For the existential quantifier we get:

$\mathbf{K}_i(\exists x A) = \mathbf{t}$ iff $\mathbf{K}'_i(A) = \mathbf{t}$ for some temporal structure $\mathbf{K}' = (\mathbf{S}, \xi', \mathbf{W})$ with $\xi'(y) \underset{|\mathbf{S}|}{=} \xi(y)$ for every y other than x.

Example. Let p be a binary predicate symbol and f a binary function symbol. Consider the formula:

$$A \equiv \exists x \circ \forall y\, p(x, f(y, a))$$

in the temporal structure $\mathbf{K} = (\mathbf{S}, \xi, \mathbf{W})$, $\mathbf{W} = \{\eta_0, \eta_1, \ldots\}$, with $|\mathbf{S}| = \mathbb{N}_0$, $\mathbf{S}(p) = <$ ("less than"), $\mathbf{S}(f) = +$ ("plus"), and $\eta_1(a) = 0$, $\eta_2(a) = 3$. Then we have:

$$\begin{aligned}
&\mathbf{K}_1(f(y, a)) = \xi(y) + 0 = \xi(y),\\
&\mathbf{K}_2(f(y, a)) = \xi(y) + 3,\\
&\mathbf{K}_1(p(x, f(y, a))) = \mathbf{t} && \Leftrightarrow \xi(x) < \xi(y),\\
&\mathbf{K}_2(p(x, f(y, a))) = \mathbf{t} && \Leftrightarrow \xi(x) < \xi(y) + 3,\\
&\mathbf{K}_0(\circ \forall y\, p(x, f(y, a))) = \mathbf{t} && \Leftrightarrow \mathbf{K}_1(\forall y\, p(x, f(y, a))) = \mathbf{t}\\
& && \Leftrightarrow \xi(x) < \xi'(y) \quad \text{for every } \xi'(y) \in \mathbb{N}_0,\\
&\mathbf{K}_1(\circ \forall y\, p(x, f(y, a))) = \mathbf{t} && \Leftrightarrow \mathbf{K}_2(\forall y\, p(x, f(y, a))) = \mathbf{t}\\
& && \Leftrightarrow \xi(x) < \xi'(y) + 3 \quad \text{for every } \xi'(y) \in \mathbb{N}_0,\\
&\mathbf{K}_0(A) = \mathbf{t} && \Leftrightarrow \text{there is } \xi''(x) \in \mathbb{N}_0 \text{ such that}\\
& && \quad \xi''(x) < \xi'(y) \quad \text{for every } \xi'(y) \in \mathbb{N}_0,\\
&\mathbf{K}_1(A) = \mathbf{t} && \Leftrightarrow \text{there is } \xi''(x) \in \mathbb{N}_0 \text{ such that}\\
& && \quad \xi''(x) < \xi'(y) + 3 \quad \text{for every } \xi'(y) \in \mathbb{N}_0.
\end{aligned}$$

So we get $\mathbf{K}_0(A) = \mathbf{f}$ since for $\xi'(y) = 0$ there is no $\xi''(x) < \xi'(y)$, and $\mathbf{K}_1(A) = \mathbf{t}$ since, e.g., $\xi''(x) = 0$ is an appropriate choice for fulfilling the condition. □

Having defined the "truth of a formula in a state" we again define the remaining semantical notions.

Definition. A formula A of $\mathscr{L}_{TP}$ is called *valid in the temporal structure* $\mathbf{K}$ ($\Vdash_{\mathbf{K}} A$) if $\mathbf{K}_i(A)=\mathbf{t}$ for every $i \in \mathbb{N}_0$. A is called *valid* ($\Vdash A$) if $\Vdash_{\mathbf{K}} A$ for every $\mathbf{K}$. *A follows from* a set $\mathscr{F}$ of closed formulas if $\Vdash_{\mathbf{K}} A$ for every $\mathbf{K}$ with $\Vdash_{\mathbf{K}} B$ for every $B \in \mathscr{F}$. A *follows from* a set $\mathscr{F}$ of formulas ($\mathscr{F} \Vdash A$) if A follows from the set of universal closures of all formulas of $\mathscr{F}$.

The only new idea is the reference to universal closures in $\mathscr{F} \Vdash A$. This comes from the fact that in the notion $\Vdash_{\mathbf{K}} A$ we only refer to one single global variable valuation ξ. (Compare this with the definition of $\Vdash_{\mathbf{S}} A$ in classical logic.)

Example. The formula $\forall x A \to A_x(t)$ is a "typical" classically valid formula (it can be taken as a basic axiom). In general, it is not valid here. Consider the case that $A \equiv x = a \to \circ(x=a)$, t is a local variable b and $\mathbf{K}$ is such that $\eta_0(a) = \eta_0(b) = \eta_1(a) \neq \eta_1(b)$. Then

$$\mathbf{K}_0(\forall x(x=a \to \circ(x=a))) = \mathbf{t} \quad \text{(because of } \eta_0(a)=\eta_1(a)\text{)},$$
$$\mathbf{K}_0(b=a)=\mathbf{t},$$
$$\mathbf{K}_1(b=a)=\mathbf{f}.$$

This amounts to

$$\mathbf{K}_0(\forall x(x=a \to \circ(x=a)) \to (b=a \to \circ(b=a))) = \mathbf{f}.$$

However, we can show that $\forall x A \to A_x(t)$ is valid in the new context of $\mathscr{L}_{TP}$ if the substitution of t for x does not create new occurrences of local variables in the scope of a temporal operator in A.

$$\begin{aligned}
\mathbf{K}_i(\forall x A)=\mathbf{t} &\Rightarrow \mathbf{K}'_i(A)=\mathbf{t} \quad \text{for every } \mathbf{K}' \text{ with } \xi'(y) \underset{|\mathbf{S}|}{=} \xi(y) \quad \text{for every } y \neq x\\
&\Rightarrow \mathbf{K}'_i(A)=\mathbf{t} \quad \text{for } \mathbf{K}' \text{ with } \xi'(x) \underset{|\mathbf{S}|}{=} \mathbf{S}^{(\xi,\eta_i)}(t) \text{ and } \xi'(y) \underset{|\mathbf{S}|}{=} \xi(y)\\
&\qquad\qquad\qquad\qquad \text{for } y \neq x\\
&\Rightarrow \mathbf{K}_i(A_x(t)) = \mathbf{K}'_i(A) \quad \text{for this } \mathbf{K}'\\
&\Rightarrow \mathbf{K}_i(A_x(t)) = \mathbf{t}. \quad \square
\end{aligned}$$

A term t with the property that its substitution for x in A does not create new occurrences of local variables in the scope of a temporal operator is called *substitutable for x in A*. If t does not contain any local variable it is substitutable for any x in any A.

9. The Formal System Σ_{TP}

We now give a formal system Σ_{TP} for first-order temporal logic.

Axioms

All axioms (taut), (ax 1)–(ax 5) of Σ_{TA} and additionally

(ax 6) $\forall x A \to A_x(t)$ if t is substitutable for x in A,

(ax 7) $\forall x \circ A \to \circ \forall x A$,

(ax 8) $A \to \circ A$ if A does not contain local variables,
(eq 1) $x = x$,
(eq 2) $x = y \to (A \to A_x(y))$ if A does not contain temporal operators.

Rules

The rules (mp), (nex), (ind) of Σ_{TA}, and additionally

(gen) $A \to B \vdash A \to \forall x B$ if there is no free occurrence of x in A.

Σ_{TP} is essentially a conglomeration of Σ_{TA} and a classical first-order system Σ_P. The only new items are (ax 7) combining one of the temporal operators with quantification and (ax 8) expressing the fact that all symbols except local variables are constantly interpreted in all states.

Theorem 9.1. (Soundness Theorem for Σ_{TP}.) *Let A be a formula and $\mathscr{F}$ a set of formulas. If $\mathscr{F} \vdash A$ then $\mathscr{F} \Vdash A$.*

Proof. The proof runs again by induction on the assumed derivation of A from $\mathscr{F}$.

1. A is an axiom: For $\mathscr{F} \Vdash A$ it suffices to show that A is valid. This was already done for the axioms of Σ_{TA} in the proof of Theorem 5.1 and for (ax 6) in the example at the end of the previous section. The remaining axioms are treated as follows:

   (ax 7): $\mathbf{K}_i(\forall x \circ A) = \mathbf{t} \Rightarrow \mathbf{K}'_{i+1}(A) = \mathbf{t}$ for every $\mathbf{K}'$ with $\xi'(y) \underset{|\mathbf{S}|}{=} \xi(y)$ for $y \not\equiv x$
   $\Rightarrow \mathbf{K}_{i+1}(\forall x A) = \mathbf{t}$
   $\Rightarrow \mathbf{K}_i(\circ \forall x A) = \mathbf{t}$.

   (ax 8): $\mathbf{K}_i(A) = \mathbf{t} \Rightarrow \mathbf{K}_i(\circ A) = \mathbf{K}_{i+1}(A) = \mathbf{t}$ since these values do not depend on η_i and η_{i+1}, respectively.

   (eq 1): $\mathbf{K}_i(x = x) = \mathbf{t}$ since $\xi(x) \underset{|\mathbf{S}|}{=} \xi(x)$ and $\eta_i(x) \underset{|\mathbf{S}|}{=} \eta_i(x)$, respectively.

   (eq 2): $\mathbf{K}_i(x = y) = \mathbf{t}$ and $\mathbf{K}_i(A) = \mathbf{t} \Rightarrow \mathbf{K}_i(A_x(y)) = \mathbf{K}_i(A) = \mathbf{t}$.

2. $A \in \mathscr{F}$: then trivially $\mathscr{F} \Vdash A$.
3. A is a conclusion of a rule of Σ_{TP}: The rules (mp), (nex) and (ind) are treated exactly as in the proof of Theorem 5.1. Consider the rule (gen). As induction hypothesis we assume $\mathscr{F} \Vdash A \to B$. Let $\mathbf{K}$ be a temporal structure, $\Vdash_{\mathbf{K}} C$ for all universal closures C of formulas of $\mathscr{F}$, and hence $\Vdash_{\mathbf{K}} A \to B$. Suppose $\mathbf{K}_i(A \to \forall x B) = \mathbf{f}$ for some $i \in \mathbb{N}_0$, i.e., $\mathbf{K}_i(A) = \mathbf{t}$ and $\mathbf{K}_i(\forall x B) = \mathbf{f}$. Then there is some $\mathbf{K}'$ with $\mathbf{K}'_i(B) = \mathbf{f}$ and $\mathbf{K}'$ differs from $\mathbf{K}$ at most in the value of ξ for x. This means also that $\Vdash_{\mathbf{K}'} C$ for every C as above and hence $\Vdash_{\mathbf{K}'} A \to B$. Since x has no free occurrence in A we have $\mathbf{K}'_i(A) = \mathbf{t}$ and hence $\mathbf{K}'_i(B) = \mathbf{t}$. This is a contradiction, so $\mathbf{K}_i(A \to \forall x B) = \mathbf{t}$ for every i, i.e., $\Vdash_{\mathbf{K}}(A \to \forall x B)$ and hence $\mathscr{F} \Vdash A \to \forall x B$. □

We can transfer to Σ_{TP} all logical laws and derived rules treated in the previous chapters since Σ_{TA} is a part of Σ_{TP}. Moreover, we may incorporate classical first-order reasoning into derivations within Σ_{TP} because of the first-order axioms and the rule (gen). So analogously to the derived rule (prop) we will use an additional rule:

(pred) $A_1, \ldots, A_n \vdash B$ whenever B is a "first-order consequence" of $A_1, \ldots, A_n$.

However, a formal definition of the notion "first-order consequence" analogous to that of "tautological consequence" used for (prop) cannot be given so easily. In particular, due to the restriction in (ax6) we have to be somewhat careful in the case when local variables are involved. The simplest precise definition of the meaning of (pred) is that B is derivable from $A_1, \ldots, A_n$ within Σ_{TP} by using only (taut), (ax6), (eq1), (eq2), (mp) and (gen). We will apply (pred) in order to abbreviate such simple classical (but sometimes tedious) derivation steps. Examples for (pred) are:

- $A \vdash \forall x A$,
- $A \rightarrow B \vdash \exists x A \rightarrow B$ if there is no free occurrence of x in B,
- $t_1 = t_2 \vdash t_2 = t_1$,
- $t_1 = t_2,\ t_2 = t_3 \vdash t_1 = t_3$.

We also want to note a first-order version of the Deduction Theorem. As in classical first-order logic one has to be a little bit careful in its formulation because of the variable condition in the rule (gen).

Theorem 9.2. (Deduction Theorem of first-order temporal logic.) *Let A, B be formulas, A closed and $\mathscr{F}$ a set of formulas. If $\mathscr{F} \cup \{A\} \vdash B$ then $\mathscr{F} \vdash \Box A \rightarrow B$.*

Proof. The proof runs again by induction on the assumed derivation of B from $\mathscr{F} \cup \{A\}$. The cases that B is an axiom or $B \in \mathscr{F} \cup \{A\}$ and the treatment of the rules (mp), (nex) and (ind) can be taken word for word from the proof of Theorem 5.3. It remains to consider rule (gen). Let $\mathscr{F} \cup \{A\} \vdash C \rightarrow D$, with x not free in C, such that $B \equiv C \rightarrow \forall x D$ is derived by (gen). By the induction hypothesis we know $\mathscr{F} \vdash \Box A \rightarrow (C \rightarrow D)$, hence $\mathscr{F} \vdash \Box A \wedge C \rightarrow D$ by (prop). Since A is closed there is no free occurrence of x in A and therefore we can apply (gen) to get $\mathscr{F} \vdash \Box A \wedge C \rightarrow \forall x D$ which yields $\mathscr{F} \vdash \Box A \rightarrow (C \rightarrow \forall x D)$ by (prop). □

Of course, the converse assertion holds again without any restrictions:

Theorem 9.3. *Let A, B be formulas, $\mathscr{F}$ a set of formulas. If $\mathscr{F} \vdash \Box A \rightarrow B$ then $\mathscr{F} \cup \{A\} \vdash B$.*

Proof. Exactly as for Theorem 5.4. □

We conclude this section by listing some more logical laws. We formally derive one of them and leave the rest as an exercise.

(T49)	$\forall x \circ A \leftrightarrow \circ \forall x A$	
(T50)	$\exists x \circ A \leftrightarrow \circ \exists x A$	
(T51)	$\forall x \Box A \leftrightarrow \Box \forall x A$	
(T52)	$\exists x \Diamond A \leftrightarrow \Diamond \exists x A$	
(T53)	$\forall x(A$ **atnext** $B) \leftrightarrow (\forall x A)$ **atnext** B	if there is no free occurrence of x in B
(T54)	$\exists x(A$ **atnext** $B) \leftrightarrow (\exists x A)$ **atnext** B	if there is no free occurrence of x in B

All of these laws state some commutativity of temporal operators with quantifiers. Notice that (ax 7) is just one direction of (T 49).

Derivation of (T51)

(1)	$\forall x \Box A \rightarrow \Box A$	(ax 6)
(2)	$\Box A \rightarrow \circ \Box A$	(prop), (ax 3)
(3)	$\forall x \Box A \rightarrow \circ \Box A$	(prop), (1), (2)
(4)	$\forall x \Box A \rightarrow \forall x \circ \Box A$	(gen), (3)
(5)	$\forall x \Box A \rightarrow \circ \forall x \Box A$	(prop), (ax 7), (4)
(6)	$\forall x \Box A \rightarrow A$	(prop), (ax 3), (1)
(7)	$\forall x \Box A \rightarrow \forall x A$	(gen), (6)
(8)	$\forall x \Box A \rightarrow \Box \forall x A$	(ind), (5), (7)
(9)	$\forall x A \rightarrow A$	(ax 6)
(10)	$\Box \forall x A \rightarrow \Box A$	(T 26), (9)
(11)	$\Box \forall x A \rightarrow \forall x \Box A$	(gen), (10)
(12)	$\forall x \Box A \leftrightarrow \Box \forall x A$	(prop), (8), (11) □

10. The Principle of Well-Founded Orderings

We now come back to a question raised in Section 7: is there any general proof principle for proving formulas of the kind:

$$A \rightarrow \Diamond B?$$

We have already argued that we should not expect to get such a principle from the formal system (or better: the logic alone). Let us consider the situation somewhat more closely. Suppose, in some state η_i, A is true and B is true in η_{i+n}, i.e., $A \rightarrow \Diamond B$ is true in η_i:

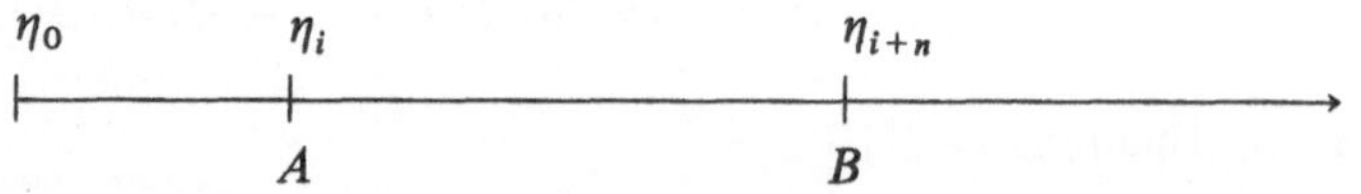

One possible way to prove $A \rightarrow \Diamond B$ is to find a chain of intermediate assertions $A_1, \ldots, A_{n-1}$ such that

$$A \rightarrow \circ A_1,$$
$$A_1 \rightarrow \circ A_2,$$
$$\vdots$$
$$A_{n-1} \rightarrow \circ B$$

can be proved. This idea is best formalized by the following two rules:

(som) $A \rightarrow \circ B \vdash A \rightarrow \Diamond B$,
(chain) $A \rightarrow \Diamond B,\ B \rightarrow \Diamond C \vdash A \rightarrow \Diamond C$,

which in fact are useful in many cases and can easily be derived even in Σ_{TA}: (som) follows directly from (ax 3) and (chain) follows with (T 15) and (T 35).

However, the problem with this principle is that the "distance" n between A and B need not be constant at all for all states where we have to check the truth of $A \rightarrow \Diamond B$. If A is true in some other state η_j, B might become true in state η_{j+m} with $m \neq n$. In general, n will vary over $\mathbb{N}_0$, and since n directly enters into the length of a proof according to (som) and (chain), it is clear that these rules cannot capture full generality. What could help is a kind of induction principle just over this "distance" n.

Once more, such a principle cannot be taken from Σ_{TA} or Σ_{TP}; it comes from "outside". But the linguistic means of $\mathscr{L}_{TP}$ at least allow one to formulate such a principle. Suppose the formula A contains a free global variable z ranging over $\mathbb{N}_0$ (we display this by $A(z)$) and we are able to prove:

$$A(z) \rightarrow \circ A(z-1)$$

where $A(z-1)$ denotes $A_z(z-1)$. By induction on z it is informally clear then that:

$$A(z) \rightarrow \Diamond A(0)$$

holds, so we only have to know $A(0) \rightarrow B$ in order to find $A \rightarrow \Diamond B$. The variable z represents just the "distance" n above.

Let us now formalize this idea. Instead of $\mathbb{N}_0$ as the range of z we want to permit any set allowing such an inductive argument. This leads to the following definition.

Definition. A relation $\preccurlyeq$ on a set Z is called a *well-founded ordering* if the following four conditions hold (for arbitrary $z, z', z'' \in Z$):

i) $z \preccurlyeq z$,
ii) $z \preccurlyeq z'$ and $z' \preccurlyeq z'' \Rightarrow z \preccurlyeq z''$,
iii) $z \preccurlyeq z'$ and $z' \preccurlyeq z \Rightarrow z = z'$,
iv) there is no infinite subset $\{z_0, z_1, z_2, \ldots\}$ of Z such that

$$z_0 \succ z_1 \succ z_2 \succ \ldots.$$

Notice that i)–iii) express that $\preccurlyeq$ is a *partial ordering*. $z \succ z'$, also written $z' \prec z$, means $z' \preccurlyeq z$ and $z' \neq z$.

A well-founded ordering $\preccurlyeq$ on a set Z gives rise to the so-called

Transfinite induction principle
If for some property $\mathfrak{p}$ *and every* $z \in Z$, $\mathfrak{p}(z)$ *(i.e., "z has property* $\mathfrak{p}$*") can be proved under the assumption that* $\mathfrak{p}(z')$ *holds for every* $z' \prec z$, *then* $\mathfrak{p}(z)$ *holds for all* $z \in Z$.

Proof. Assume there is some z_0 such that not $\mathfrak{p}(z_0)$. Then there must be $z_1 \prec z_0$ such that not $\mathfrak{p}(z_1)$ since otherwise $\mathfrak{p}(z_0)$ could be proved. This argument can be applied infinitely often, yielding an infinite sequence $z_0 \succ z_1 \succ z_2 \succ \ldots$, hence a contradiction. □

Of course, the usual ordering $\leq$ is a well-founded ordering on $\mathbb{N}_0$. Transfinite induction is then just the usual mathematical induction.

Suppose now a language $\mathscr{L}_{TP}$ to contain a special binary predicate symbol $\preccurlyeq$ and some distinguished global variables (denoted $z, z', z_1, z_2, \ldots$) on which $\preccurlyeq$ can be applied (i.e., $z \preccurlyeq z'$, etc., are atomic formulas of $\mathscr{L}_{TP}$), and define $z \prec z'$ to be $z \preccurlyeq z' \wedge z \neq z'$. Consider, moreover, the situation where every temporal structure $(\mathbf{S}, \xi, \mathbf{W})$ has the property that $|\mathbf{S}|$ contains some set Z such that:

- $\xi(z), \xi(z'), \ldots \in Z$,
- $\mathbf{S}(\preccurlyeq)$ is a well-founded ordering on Z.

In this case we call $\mathscr{L}_{TP}$ a *language with well-founded ordering* and denote it by $\mathscr{L}_{TP}^{wf}$.

According to the second condition above, the predicate symbol $\preccurlyeq$ is interpreted as a well-founded ordering. So it should be clear from the whole discussion that the following formulas are valid in the situation described and may be taken as additional axioms in every $\mathscr{L}_{TP}^{wf}$:

(po 1)	$z \preccurlyeq z$,
(po 2)	$z_1 \preccurlyeq z_2 \wedge z_2 \preccurlyeq z_3 \rightarrow z_1 \preccurlyeq z_3$,
(po 3)	$z_1 \preccurlyeq z_2 \wedge z_2 \preccurlyeq z_1 \rightarrow z_1 = z_2$,
(ti)	$\forall z(\forall z'(z' \prec z \rightarrow A(z')) \rightarrow A(z)) \rightarrow A(z)$.

(ti) is a formalization of the transfinite induction principle.

In any $\mathscr{L}_{TP}^{wf}$ we are able now to formulate and prove our proof principle in the following general form:

(wfo) $\quad A(z) \rightarrow \Diamond(B \vee \exists z'(z' \prec z \wedge A(z'))) \vdash \exists z A(z) \rightarrow \Diamond B$
if B does not contain z.

Derivation

(1)	$A(z) \rightarrow \Diamond(B \vee \exists z'(z' \prec z \wedge A(z')))$	assumption
(2)	$\Diamond A(z) \rightarrow \Diamond(B \vee \exists z'(z' \prec z \wedge A(z')))$	(T 42), (1)
(3)	$\Diamond A(z) \rightarrow \Diamond B \vee \exists z'(\Diamond(z' \prec z) \wedge \Diamond A(z'))$	(T 23), (T 29), (T 52), (2)
(4)	$\neg z' \prec z \rightarrow \circ \neg z' \prec z$	(ax 8)
(5)	$\neg z' \prec z \rightarrow \Box \neg z' \prec z$	(ind'), (4)
(6)	$\Diamond(z' \prec z) \rightarrow z' \prec z$	(prop), (T 2), (5)
(7)	$\Diamond A(z) \rightarrow \Diamond B \vee \exists z'(z' \prec z \wedge \Diamond A(z'))$	(pred), (3), (6)
(8)	$\exists z'(z' \prec z \wedge \Diamond A(z')) \wedge$ $\forall z'(z' \prec z \rightarrow (\Diamond A(z') \rightarrow \Diamond B)) \rightarrow \Diamond B$	(pred)
(9)	$\Diamond A(z) \wedge \forall z'(z' \prec z \rightarrow (\Diamond A(z') \rightarrow \Diamond B)) \rightarrow \Diamond B$	(prop), (T 15), (7), (8)
(10)	$\forall z'(z' \prec z \rightarrow (\Diamond A(z\text{-}') \rightarrow \Diamond B)) \rightarrow (\Diamond A(z) \rightarrow \Diamond B)$	(prop), (9)
(11)	$\Diamond A(z) \rightarrow \Diamond B$	(ti), (mp), (10)
(12)	$A(z) \rightarrow \Diamond A(z)$	(prop), (ax 4)
(13)	$A(z) \rightarrow \Diamond B$	(prop), (11), (12)
(14)	$\exists z A(z) \rightarrow \Diamond B$	(pred), (13) $\Box$

It should be noted that the premise in (wfo) is itself a formula of the form $C \rightarrow \Diamond D$, so this rule cannot be used without some further means. A trivial case is if the premise is given by $C \rightarrow \circ D$ applying (som). In other cases, the premise of (wfo) is to be established by other means, e.g., by applying (som) together with (chain).

We want finally to note that applying the rule (chain) turns out to be a very special case of applying (wfo). This is not surprising in the light of the informal discussion at the beginning of this section. Formally we see this fact as follows.

Suppose we are able to prove:

$$B_1 \rightarrow \Diamond B_2,\ B_2 \rightarrow \Diamond B_3,\ \ldots,\ B_{k-1} \rightarrow \Diamond B_k \qquad (k \geq 2)$$

from which we get $B_1 \rightarrow \Diamond B_k$ by applying (chain) $k-1$ times. Let $A_{k-1} \equiv B_1$, $A_{k-2} \equiv B_2, \ldots, A_1 \equiv B_{k-1}$, $B = B_k$. Then we have

$$A_{k-1} \rightarrow \Diamond A_{k-2},\ \ldots,\ A_2 \rightarrow \Diamond A_1,\ A_1 \rightarrow \Diamond B.$$

Now let

$$A(n) \equiv (n = k-1 \wedge A_{k-1}) \vee \ldots \vee (n = 1 \wedge A_1),$$

n ranging over $\mathbb{N}_0$. (Informally this means that we view A_i as $A(i)$.) Then we have:

$$A(n) \rightarrow \Diamond (B \vee \exists m (m < n \wedge A(m)))$$

and we get $\exists n A(n) \rightarrow \Diamond B$ by (wfo), hence $B_1 \rightarrow \Diamond B$ because of $B_1 \rightarrow \exists n A(n)$.

11. Additional Propositional Variables

One major difference between propositional and first-order logic is that atomic formulas in propositional logic are basic, not-further-analyzed entities, whereas any atomic formula in first-order logic has the linguistic structure of an application of a predicate (symbol) to terms. Terms may contain subject variables and in analogy to these we may view atomic formulas in the first sense as *propositional variables* ranging over the truth values $\{\mathbf{f}, \mathbf{t}\}$.

In certain applications – and we will find one of this kind in the next chapter – it is desirable to have a first-order language with additional propositional variables, i.e., a language in which additional atomic formulas exist which are not of the standard form $p(t_1, \ldots, t_n)$ and are evaluated in $\{\mathbf{f}, \mathbf{t}\}$ by some additional assignment. Such an extension of a first-order language is without any problems and we sketch it briefly for the case $\mathscr{L}_{\mathrm{TP}}$. (With regard to later applications we consider only "local" propositional variables.)

Let $\mathscr{V}$ be a denumerable set of *propositional variables*. A *first-order temporal language* $\mathscr{L}_{\mathrm{TP}}^{+}$ *augmented by* $\mathscr{V}$ is a language $\mathscr{L}_{\mathrm{TP}}$ with the additional grammatical rule that any $v \in \mathscr{V}$ is an atomic formula too.

In order to define the semantics of $\mathscr{L}_{\mathrm{TP}}^{+}$ we have to extend the notion of temporal structure $\mathbf{K} = (\mathbf{S}, \xi, \mathbf{W})$: $\mathbf{S}$ and ξ are as in the case of $\mathscr{L}_{\mathrm{TP}}$ but any state $\eta_i \in \mathbf{W}$ is now of the form:

$$\eta_i = (\eta_i^{(1)}, \eta_i^{(2)})$$

where $\eta_i^{(1)}$ is again a local (subject) variable valuation with respect to $\mathbf{S}$ and $\eta_i^{(2)}$ is a Boolean valuation, i.e., a mapping:

$$\eta_i^{(2)}: \mathscr{V} \rightarrow \{\mathbf{f}, \mathbf{t}\}.$$

Again we find a truth value $\mathbf{S}^{(\xi, \eta_i)}(A)$ for every atomic formula by applying the definitions noted in Section 8 for the usual first-order atomic formulas (but writing $\eta_i^{(1)}$ instead of η_i) and defining:

$$\mathbf{S}^{(\xi, \eta_i)}(v) = \eta_i^{(2)}(v)$$

for the new ones. The rest of the semantic definitions are the same as for $\mathscr{L}_{\mathrm{TP}}$.

Example. Consider the following formula A of $\mathscr{L}_{\mathrm{TP}}^{+}$:

$$A \equiv \circ \neg v_1 \vee (\forall x\, p(x, a) \rightarrow v_2)$$

($v_1, v_2 \in \mathscr{V}$, p predicate symbol) and the temporal structure $\mathbf{K} = (\mathbf{S}, \xi, \mathbf{W})$ with $|\mathbf{S}| = \mathbf{N}_0$, $\mathbf{S}(p) = \geq$ ("greater or equal") and $\mathbf{W}$ given according to the following matrix:

	η_0	η_1	η_2	η_3	
a	7	0	1	3 ...	} values of $\eta_i^{(1)}$
v_1	**f**	**t**	**f**	**f** ...	} values of $\eta_i^{(2)}$.
v_2	**f**	**f**	**t**	**f** ...	

We calculate:

$$\mathbf{K}_0(p(x, a)) = \mathbf{t} \Leftrightarrow \xi(x) \geq 7,$$
$$\mathbf{K}_0(\forall x\, p(x, a)) = \mathbf{f},$$
$$\mathbf{K}_0(v_2) = \eta_0^{(2)}(v_2) = \mathbf{f},$$
$$\mathbf{K}_0(\forall x\, p(x, a) \rightarrow v_2) = \mathbf{t}.$$

Hence, we find $\mathbf{K}_0(A) = \mathbf{t}$. We also have $\mathbf{K}_1(A) = \mathbf{t}$ because:

$$\mathbf{K}_2(v_1) = \eta_2^{(2)}(v_1) = \mathbf{f},$$

and therefore:

$$\mathbf{K}_1(\circ \neg v_1) = \mathbf{K}_2(\neg v_1) = \mathbf{t}. \quad \square$$

Of course, any formula of $\mathscr{L}_{\mathrm{TP}}$ is a formula of $\mathscr{L}_{\mathrm{TP}}^{+}$ and if it is valid according to the semantics of $\mathscr{L}_{\mathrm{TP}}$ it is also valid in $\mathscr{L}_{\mathrm{TP}}^{+}$. The converse is not quite correct. We have new valid formulas in $\mathscr{L}_{\mathrm{TP}}^{+}$, for example, the formula:

$$v \rightarrow v$$

which is not contained in $\mathscr{L}_{\mathrm{TP}}$. However, any such new formula must obviously fall under a *scheme* of formulas, in the example the scheme:

$$A \rightarrow A$$

which is "valid" in $\mathscr{L}_{\mathrm{TP}}$. More precisely, a formula of $\mathscr{L}_{\mathrm{TP}}^{+}$ containing propositional variables can only be valid if it is tautologically valid (extending this notion to $\mathscr{L}_{\mathrm{TP}}^{+}$). An analogous consideration holds for the notion "follows from".

This also shows that the formal system Σ_{TP} is good for $\mathscr{L}_{TP}^{+}$ too, if we modify (ax 8) in the following obvious way:

(ax 8$^+$) $A \rightarrow \circ A$ if A contains neither local variables nor propositional variables.

It is sound with respect to the semantics of $\mathscr{L}_{TP}^{+}$. Moreover, all further investigations and results of Sections 9 and 10 remain valid for $\mathscr{L}_{TP}^{+}$.

With these remarks we conclude our purely logical theory. To summarize, we have developed a "hierarchy" of logical languages and systems. Including classical logic we may depict this hierarchy as follows:

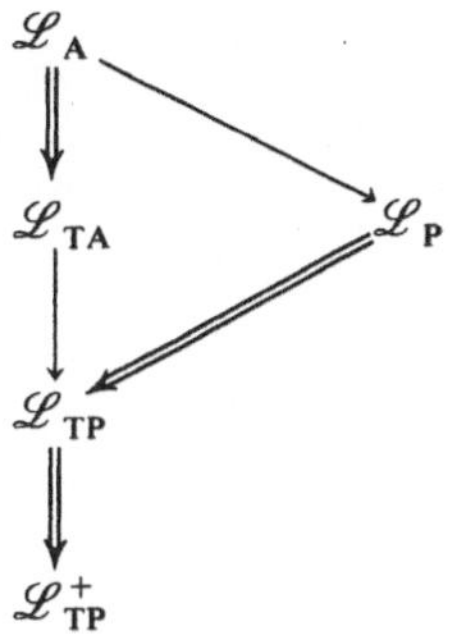

An arrow relation $\mathscr{L}_1 \rightarrow \mathscr{L}_2$ means that $\mathscr{L}_1$ is a "sublogic" of $\mathscr{L}_2$. A double arrow $\mathscr{L}_1 \Rightarrow \mathscr{L}_2$ means that $\mathscr{L}_1$ is even, moreover, a sublanguage of $\mathscr{L}_2$.

Chapter IV

Temporal Semantics of Programs

12. Programs

We want to apply temporal logic to the description of (the computational behaviour of) programs and their verification. For the sake of technical conciseness we restrict the programs to a fixed (somewhat abstract) syntactic form.

A program Π consists of a number of *parallel components* $\Pi_1, \ldots, \Pi_p$ which are thought to be executed in parallel and each of which is a sequential "while-program". In the case $p=1$, Π is itself a sequential program. The parallel components can either all be *cyclic* (not intended to terminate) or *non-cyclic* (intended to terminate). In order to synchronize parallel components, a special kind of statement:

await ... **then** ...

is allowed. A program in this sense is "structured" (without goto's); nevertheless, we label any statement (except under an **await**) by a (unique) label. The reason is that we want to be able to identify the single locations of computation in a concrete execution of the program.

According to these ideas we make the following formal definitions.

Let $\mathscr{L}_\mathrm{P}$ be a (classical) first-order language and $\mathscr{E}$ a set of *elementary statements* about which we make no further assumptions at the moment. We define $\mathscr{S}$ to be the set of *synchronization statements* of the form:

await B **then** α or
await B

where B is a formula of $\mathscr{L}_\mathrm{P}$ and $\alpha \in \mathscr{E}$. Finally, let $\mathscr{A} = \mathscr{E} \cup \mathscr{S}$.

A *program over* $\mathscr{L}_\mathrm{P}$ *and* $\mathscr{A}$ is a string of the form

initial R;
cobegin $\Pi_1 \parallel \ldots \parallel \Pi_p$ **coend**

where R is a formula of $\mathscr{L}_\mathrm{P}$ called the *initialization condition*, $p \geq 1$, and every Π_i, $1 \leq i \leq p$, is either a ⟨cyclic parallel component⟩ or a ⟨non-cyclic parallel component⟩ according to the following syntactical rules:

⟨cyclic parallel component⟩ : : = **loop** ⟨statement sequence⟩ **end**
⟨non-cyclic parallel component⟩ : : = ⟨statement sequence⟩; ⟨label⟩: **stop**
⟨statement sequence⟩ : : = ⟨statement⟩ | ⟨statement⟩; ⟨statement sequence⟩
⟨statement⟩ : : = ⟨label⟩: ⟨unlabelled statement⟩

⟨unlabelled statement⟩: : = ⟨element of $\mathscr{A}$⟩ |
  if ⟨condition⟩ **then** ⟨statement sequence⟩
    else ⟨statement sequence⟩ **fi** |
  if ⟨condition⟩ **then** ⟨statement sequence⟩ **fi** |
  while ⟨condition⟩ **do** ⟨statement sequence⟩ **od**

⟨condition⟩: : = ⟨formula of $\mathscr{L}_P$⟩

A ⟨label⟩ is an element of some given set of labels, in our programs mostly $\{\alpha_0, \alpha_1, \alpha_2, \ldots, \beta_0, \beta_1, \beta_2, \ldots\}$. In any program the labels must fulfill the following:

Label condition. No two labels occurring in a program are equal.

A program is called *cyclic* if it consists of ⟨cyclic parallel components⟩ and *non-cyclic* in the other case.

Whenever $p=1$ we will omit the parentheses **cobegin** and **coend**, and throughout the rest of this volume we will use the following notation:

$\alpha_0^{(i)}$: label of the first statement in component Π_i (*start label*),
$\alpha_e^{(i)}$: label of **stop** in the non-cyclic component Π_i (*end label*),
$\mathscr{M}_{\Pi_i}$: set of labels occurring in component Π_i,

$$\mathscr{M}_\Pi = \bigcup_{i=1}^{p} \mathscr{M}_{\Pi_i},$$

$$\bar{\mathscr{M}}_{\Pi_i} = \mathscr{M}_{\Pi_i} \setminus \{\alpha_e^{(i)}\},$$

$$\bar{\mathscr{M}}_\Pi = \mathscr{M}_\Pi \setminus \{\alpha_e^{(1)}, \ldots, \alpha_e^{(p)}\} = \bigcup_{i=1}^{p} \bar{\mathscr{M}}_{\Pi_i}.$$

Example. Assuming that $\mathscr{L}_P$ contains the language of Peano arithmetic and $\mathscr{E}$ contains usual assignment statements, the following is a (non-cyclic) program:

$\Pi \equiv$ **initial** $a=0 \wedge b=1$;
  cobegin α_0: $b:=b \times b$;
    α_1: **await** $a \neq 0$ **then** $a:=1$;
    α_2: **if** $a<b$ **then** α_3: $b:=b-a$ **fi**;
    α_4: $b:=2 \times b$;
    α_5: **stop**
    ‖
    β_0: **while** $b=1$ **do** β_1: $a:=a+2$;
        β_2: $b:=2 \times b$ **od**;
    β_3: **stop**
  coend

Here we have $\alpha_0^{(1)} = \alpha_0$, $\alpha_0^{(2)} = \beta_0$, $\alpha_e^{(1)} = \alpha_5$, $\alpha_e^{(2)} = \beta_3$,

$\mathscr{M}_{\Pi_1} = \{\alpha_0, \alpha_1, \alpha_2, \alpha_3, \alpha_4, \alpha_5\}$,
$\mathscr{M}_{\Pi_2} = \{\beta_0, \beta_1, \beta_2, \beta_3\}$,
$\bar{\mathscr{M}}_{\Pi_1} = \{\alpha_0, \alpha_1, \alpha_2, \alpha_3, \alpha_4\}$,
$\bar{\mathscr{M}}_{\Pi_2} = \{\beta_0, \beta_1, \beta_2\}$,
$\mathscr{M}_\Pi = \{\alpha_0, \ldots, \alpha_5, \beta_0, \ldots, \beta_3\}$,
$\bar{\mathscr{M}}_\Pi = \{\alpha_0, \ldots, \alpha_4, \beta_0, \ldots, \beta_2\}$. □

Let Π_i and Π_j be two different parallel components of a program Π. We have already noted that Π_i and Π_j are thought to be executed "in parallel" (or "concurrently"). Suppose that during such an execution, α and β are the labels of (elementary) statements in Π_i and Π_j, respectively, which are to be executed next. We model the parallel execution of α and β by considering arbitrary sequentializations of α and β, i.e., by considering the possible computations:

$$\alpha\!:\sim;\beta\!:\sim, \quad \text{and}$$
$$\beta\!:\sim;\alpha\!:\sim.$$

This model of computation is called the *interleaving model.* Applied to the whole program Π it means the following.

The execution of Π takes place in discrete steps. We may think of a *scheduler* which, at every step, schedules some parallel component which executes its next (*indivisible*) *action.* Such a one-step action is:

- a statement of $\mathscr{A}$, or
- a test of the condition in an **if** statement together with the entry into the respective branch, or
- a test of the condition in a **while** statement together with the respective entry into or exit from the loop.

The "flow of control" in **if** and **while** statements is as usual. The statement sequence between some **loop** and **end** is to be repeated continuously. For **await** statements we have an additional rule: if **await** B (**then** a) is the next action to be executed in some Π_i then Π_i may be scheduled for the next step only if B holds at this moment. Executing **await** B **then** a means then executing a. Executing **await** B has no further effect. Notice that testing B and executing a (if at all) is *one* indivisible action.

Example. A possible sequence of steps executing the sample program above is described by the following table.

step	action	values of a,	b after the action
		initially: 0	1
1	α_0: $b:=b\times b$	0	1
2	β_0: entry into loop body	0	1
3	β_1: $a:=a+2$	2	1
4	α_1: **await** $a\neq 0$ **then** $a:=1$	1	1
5	β_2: $b:=2\times b$	1	2
6	α_2: entry into **then** branch	1	2
7	α_3: $b:=b-a$	1	1
8	β_0: entry into loop body	1	1
9	α_4: $b:=2\times b$	1	2
10	β_1: $a:=a+2$	3	2
11	β_2: $b:=2\times b$	3	4
12	β_0: exit from loop	3	4

After the 12th step the two components are at α_5 and β_3, respectively, which means that the execution terminates. (We think of a "statement" α: **stop** not as an action to be executed but only to be "reached".)

Because of the **await** rule the following beginning of an execution of the same program is not possible:

1	α_0: $b := b \times b$
2	α_1: **await** $a \neq 0$ **then** $a := 1$

since after the first step the value of a is not $\neq 0$. □

Because of the **await** rule some executions may end in a *deadlock*. This is a situation where, in each of those parallel components which are not yet terminated, the next action would be an **await** statement and none of the respective conditions holds, hence none of the components can be scheduled.

Example. Consider the program:

```
initial a=0 ∧ b=0;
cobegin loop α0: a:=a+1;
             α1: await b≠0;
             α2: a:=a+3 end
        ||
        loop β0: a:=2×a;
             β1: await a≠1;
             β2: b:=b+1 end
coend
```

and assume the following beginning of an execution:

step	action	a	b
	initially:	0	0
1	β_0: $a := 2 \times a$	0	0
2	α_0: $a := a + 1$	1	0

After these two actions deadlock arises because neither α_1 nor β_1 can be executed next. Of course, another scheduling might avoid a deadlock, e.g.,

step	action	a	b
	initially:	0	0
1	α_0: $a := a + 1$	1	0
2	β_0: $a := 2 \times a$	2	0
3	β_1: **await** $a \neq 1$	2	0
4	β_2: $b := b + 1$	2	1
5	α_1: **await** $b \neq 0$	2	1
⋮	and so on without any deadlock		

At first glance, the interleaving model of computation seems to be much too simplifying. Consider the two actions:

$$\alpha: a := a+1 \qquad \beta: a := a-1$$

A really parallel execution of α and β at some moment where, say, $a=0$ holds could produce different results depending on the different "speeds" of the single machine actions involved by α and β. In the interleaving model α and β are executed one after the other always yielding the value $a=0$.

Obviously, the adulteration comes from the fact that we view $a := a+1$ and $a := a-1$ as indivisible or smallest action units. We would get a more precise picture if we decompose these statements into one-address actions, e.g.,

$$\begin{array}{llll} \alpha: & \alpha_0: \mathtt{load}\ a; & \beta: & \beta_0: \mathtt{load}\ a; \\ & \alpha_1: \mathtt{add}\ 1; & & \beta_1: \mathtt{add}\ -1; \\ & \alpha_2: \mathtt{store}\ a & & \beta_2: \mathtt{store}\ a \end{array}$$

Assuming that these single actions can be carried out indivisibly (in effect) different speeds would lead to results $a=0$, $a=1$, or $a=-1$. However, interleaving α_0, α_1, α_2, β_0, β_1, β_2 in arbitrary ways also yields these results. Any single interleaving is just modelling one particular "relationship" of the single speeds. Thus, at an appropriate level of "atomicity" interleaving can be viewed as a sufficient model of parallel execution.

Although in many cases, some $a := a+1$ might not be an appropriate level of atomicity we will use such assignments or even more complex statements as elementary statements in subsequent programs. The reason is that any coherent block of statements, e.g., α_0, α_1, α_2 above, can be made indivisible by some synchronization. We will see this in Section 16 under the catchword "mutual exclusion" and will come back to this discussion there.

13. Execution Sequences of Programs

We now want to formalize the "operational semantics" of a program Π (over some $\mathscr{L}_P$, $\mathscr{A}$) described in the previous section.

Let $\mathscr{F}(\mathscr{L}_P)$ be the set of all formulas of $\mathscr{L}_P$. We first associate inductively with any statement sequence ψ (over $\mathscr{L}_P$, $\mathscr{A}$) the following three entities:

- entry $(\psi) \in \mathscr{M}_\psi$,
- $\mathrm{trans}(\psi) \subset \mathscr{M}_\psi \times \mathscr{F}(\mathscr{L}_P) \times \mathscr{M}_\psi$,
- $\mathrm{exits}(\psi) \subset \mathscr{M}_\psi \times \mathscr{F}(\mathscr{L}_P)$.

($\mathscr{M}_\psi$ is the set of labels occurring in ψ.)

1. $\psi \equiv \alpha$: $\langle$element of $\mathscr{E}\rangle$:
   $\mathrm{entry}(\psi) = \alpha$,
   $\mathrm{trans}(\psi) = \emptyset$,
   $\mathrm{exits}(\psi) = \{(\alpha, \mathbf{true})\}$.

2. $\psi \equiv \alpha$: **await** B **then** a or $\psi \equiv \alpha$: **await** B:
   entry$(\psi) = \alpha$,
   trans$(\psi) = \emptyset$,
   exits$(\psi) = \{(\alpha, B)\}$.

3. $\psi \equiv \alpha$: **if** B **then** ψ_1 **else** ψ_2 **fi**:
   entry$(\psi) = \alpha$,
   trans$(\psi) = $ trans$(\psi_1) \cup$ trans$(\psi_2) \cup \{(\alpha, B, \text{entry}(\psi_1)), (\alpha, \neg B, \text{entry}(\psi_2))\}$,
   exits$(\psi) = $ exits$(\psi_1) \cup$ exits(ψ_2).

4. $\psi \equiv \alpha$: **if** B **then** ψ_1 **fi**:
   entry$(\psi) = \alpha$,
   trans$(\psi) = $ trans$(\psi_1) \cup \{(\alpha, B, \text{entry}(\psi_1))\}$,
   exits$(\psi) = $ exits$(\psi_1) \cup \{(\alpha, \neg B)\}$.

5. $\psi \equiv \alpha$: **while** B **do** ψ_1 **od**:
   entry$(\psi) = \alpha$,
   trans$(\psi) = $ trans$(\psi_1) \cup \{(\alpha, B, \text{entry}(\psi_1))\} \cup \{(\beta, C, \alpha) \mid (\beta, C) \in \text{exits}(\psi_1)\}$,
   exits$(\psi) = \{(\alpha, \neg B)\}$.

6. $\psi \equiv \alpha$: ψ_1 ; ψ_2 (ψ_1 unlabelled statement, ψ_2 statement sequence):
   entry$(\psi) = \alpha$,
   trans$(\psi) = $ trans$(\psi_1) \cup$ trans$(\psi_2) \cup \{(\beta, C, \text{entry}(\psi_2)) \mid (\beta, C) \in \text{exits}(\psi_1)\}$,
   exits$(\psi) = $ exits(ψ_2).

Now let Π_i be some parallel component of Π. We define the set trans$(\Pi_i) \subset \mathcal{M}_{\Pi_i} \times \mathcal{F}(\mathcal{L}_P) \times \mathcal{M}_{\Pi_i}$ of *transitions* of Π_i:

1. $\Pi_i \equiv \psi; \alpha$: **stop** ($\psi$ statement sequence):
   trans$(\Pi_i) = $ trans$(\psi) \cup \{(\beta, C, \alpha) \mid (\beta, C) \in \text{exits}(\psi)\}$.

2. $\Pi_i \equiv$ **loop** ψ **end** (ψ statement sequence):
   trans$(\Pi_i) = $ trans$(\psi) \cup \{(\beta, C, \text{entry}(\psi)) \mid (\beta, C) \in \text{exits}(\psi)\}$.

Example. Consider the parallel component

$\Pi_i \equiv \alpha_0 : a := a+1;$
  α_1 : **while** $b=0$ **do** $\alpha_2 : b := b-1;$
    $\alpha_3 : a := a \times a;$
    α_4 : **await** $a < c$ **then** $a := c - a$ **od**;
  α_5 : **stop**

Applying the above definitions we get:

$$\text{trans}(\Pi_i) = \{(\alpha_0, \mathbf{true}, \alpha_1), (\alpha_1, b=0, \alpha_2), (\alpha_2, \mathbf{true}, \alpha_3), (\alpha_3, \mathbf{true}, \alpha_4), (\alpha_4, a<c, \alpha_1), (\alpha_1, b \neq 0, \alpha_5)\}$$ □

Informally, $(\alpha, C, \beta) \in$ trans(Π_i) means: In Π_i, execution can proceed in one step from α to β if C holds. Furthermore we associate with every $\alpha \in \bar{\mathcal{M}}_\Pi$ a formula E_α of $\mathcal{L}_P$, called the *full exit condition* of α:

$$E_\alpha = \begin{cases} B, & \text{if } \alpha \text{ is the label of a statement } \textbf{await } B \text{ (}\textbf{then } a\text{) in } \Pi, \\ \textbf{true}, & \text{otherwise.} \end{cases}$$

Obviously, E_α expresses the condition under which the action labelled by α may be executed.

In the syntactic definition of our programs the precise form of the elementary statements was left open. We do not want to restrict this freedom; however, we want to assume for every program that the statements "change variables" (as, e.g., assignments). We will refer to those variables changed by a particular program as the *program variables* of this program. Furthermore, we assume the variables of $\mathscr{L}_P$ to be again partitioned into local and global ones (quantification only over the latter) and establish that program variables are always local and all other variables used in the context of the program are global.

Now let

$$\Pi \equiv \texttt{initial}\ R;$$
$$\texttt{cobegin}\ \Pi_1 \parallel \ldots \parallel \Pi_p\ \texttt{coend}$$

be a program over $\mathscr{L}_P$ and $\mathscr{A}$, and let $\mathbf{S}$ be a structure for $\mathscr{L}_P$. A *program state* of Π (w.r.t. $\mathbf{S}$) is a $(p+2)$-tuple $\eta = (\mu, \lambda_1, \ldots, \lambda_p, \kappa)$ where:

- μ assigns a value $\mu(a) \in |\mathbf{S}|$ to every program variable a (*memory state*),
- $\lambda_i \in \mathscr{M}_{\Pi_i}$ for every $i = 1, \ldots, p$,
- $\kappa \in \{0, 1, \ldots, p\}$.

The informal meaning of the elements in a program state is as follows:

- $\mu(a)$ is the "value" of a.
- λ_i is the label of the action to be executed next in Π_i.
- $\kappa \neq 0$ means that Π_κ is the next scheduled parallel component (this implies that λ_κ is the label of the next executed statement).
- $\kappa = 0$ means that none of the parallel components may be scheduled because of the **await** rule or since Π has reached all its end labels. In this case we call η the *terminal state.*

The part μ of a program state η is just a (partial) variable valuation in the sense of classical first-order logic. Therefore, together with a global valuation ξ we get in the usual way the mapping $\mathbf{S}^{(\xi,\mu)}$ associating a truth value with every formula of $\mathscr{L}_P$. $\mathbf{S}^{(\xi,\mu)}(A)$ is the "truth value of A in state η" (under $\mathbf{S}$ and ξ).

Now we are able to formalize the concept of possible computations of Π in the interleaving model.

Definition. An *execution sequence of* Π (w.r.t. $\mathbf{S}$, ξ) is an infinite sequence $\mathbf{W}_\Pi = \{\eta_0, \eta_1, \eta_2, \ldots\}$ of program states of Π (w.r.t. $\mathbf{S}$) with the following properties:

- $\eta_0 = (\mu_0, \alpha_0^{(1)}, \ldots, \alpha_0^{(p)}, \kappa_0)$ and $\mathbf{S}^{(\xi,\mu_0)}(R) = \mathbf{t}$.
- If $\eta_j = (\mu, \lambda_1, \ldots, \lambda_i, \ldots, \lambda_p, i)$ then $\eta_{j+1} = (\mu', \lambda_1, \ldots, \lambda_i', \ldots, \lambda_p, \kappa')$ and trans(Π_i) contains an element $(\lambda_i, C, \lambda_i')$ and $\mathbf{S}^{(\xi,\mu)}(C) = \mathbf{t}$. If, moreover, λ_i is the label of a statement $\notin \mathscr{A}$ then $\mu' = \mu$.
- If $\eta_j = (\mu, \lambda_1, \ldots, \lambda_p, 0)$ then $\eta_{j+1} = \eta_j$ and for every $i = 1, \ldots, p$ either $\mathbf{S}^{(\xi,\mu)}(E_{\lambda_i}) = \mathbf{f}$ or $\lambda_i = \alpha_e^{(i)}$.

This definition reflects our discussion in the previous section. Notice that if a terminal state is reached the program remains in this state forever. We could also cut the sequence in such a state but it is convenient to have an infinite sequence in all cases. A more important observation is that – in general – we state no relationship between μ and μ' in the case of going from η_j to η_{j+1} according to a transition of some Π_i. This means that an execution sequence in the above sense does not contain the information about the actual changes of the variables caused by single actions. (We only state that actions not contained in $\mathscr{A}$ do not change any variable.) Again, we maintain the freedom of not yet being forced to fix what elementary actions are really performed. We will come back to this in the next section.

Example. Consider the program:

$$
\begin{array}{l}
\Pi \equiv \texttt{initial}\, a=0 \wedge b=0; \\
\quad \texttt{cobegin loop}\, \alpha_0\colon \sim; \\
\qquad\qquad \alpha_1\colon \texttt{await}\, b \neq 0; \\
\qquad\qquad \alpha_2\colon \sim \qquad\qquad \texttt{end} \\
\qquad \| \\
\qquad\quad \texttt{loop}\, \beta_0\colon \sim; \\
\qquad\qquad \beta_1\colon \texttt{await}\, a \neq 1; \\
\qquad\qquad \beta_2\colon \sim \qquad\qquad \texttt{end} \\
\quad \texttt{coend}
\end{array}
$$

with arbitrary actions under α_0, α_2, β_0, β_2. Let $|\mathbf{S}|=\mathbb{N}_0$. A possible execution sequence is:

$$
\begin{array}{lll}
\eta_0=(\mu_0, \alpha_0, \beta_0, 2) & \text{with} \quad \mu_0(a)=0, & \mu_0(b)=0, \\
\eta_1=(\mu_1, \alpha_0, \beta_1, 1), & & \\
\eta_2=(\mu_2, \alpha_1, \beta_1, 1) & \text{with} \quad \mu_2(b)\neq 0, & \\
\eta_3=(\mu_3, \alpha_2, \beta_1, 1), & & \\
\eta_4=(\mu_4, \alpha_0, \beta_1, 2) & \text{with} \quad \mu_4(a)\neq 1, & \\
\eta_5=(\mu_5, \alpha_0, \beta_2, 1), & & \\
\eta_6=(\mu_6, \alpha_1, \beta_2, 2), & & \\
\eta_7=(\mu_7, \alpha_1, \beta_0, 2), & & \\
\eta_8=(\mu_8, \alpha_1, \beta_1, 0) & \text{with} \quad \mu_8(a)=1, & \mu_8(b)=0, \\
\eta_9=\eta_8 & & \\
\vdots & & \\
\text{etc.} \quad \square & &
\end{array}
$$

14. Program Axioms

Temporal logic, as we have defined it, treats formulas over a sequence of states. We may put this another way. Considering single states η_i, η_j, etc., we are able to "describe" each one of them by non-temporal formulas and certain relationships between them by temporal connectives.

Let us first pursue the non-temporal aspect. Take a first-order language $\mathscr{L}_{TP}$. A state η_i determines the values of subject variables, e.g., $\eta_i(a)=3$ and $\eta_i(b)=7$.

Every formula valid under this valuation is a "description" of η_i, e.g.,

$$a<b,$$
$$a=3,$$
$$b=7 \wedge a>0.$$

Our goal is to describe program states in this sense. Obviously, a "pure" first-order $\mathscr{L}_{TP}$ is not sufficient for this purpose. Of course, we can express the information given by the memory state μ in the way just indicated, but there is no means in $\mathscr{L}_{TP}$ to express the information about the "location counters" $\lambda_1, \ldots, \lambda_p$, κ in a program state. However, there is a simple possibility to achieve this latter expressibility as well: we augment $\mathscr{L}_{TP}$ by some suitable propositional variables.

Let

$\Pi \equiv$ **initial** R;
  cobegin $\Pi_1 \| \ldots \| \Pi_p$ **coend**

be a program over a first-order language $\mathscr{L}_P$ and some $\mathscr{A}$. The language $\mathscr{L}_{TP\Pi}$ of *temporal logic of* Π is the first-order language $\mathscr{L}_{TP}$ with kernel $\mathscr{L}_P$ augmented by the set of the following propositional variables:

- λ for every $\lambda \in \bar{\mathscr{M}}_\Pi$,
- at λ for every $\lambda \in \mathscr{M}_\Pi$.

So every $\lambda \in \bar{\mathscr{M}}_\Pi$ is now also an (atomic) formula informally meaning:

"The action λ is executed (next)".

Moreover, we have atomic formulas at λ with the informal meaning:

"The action λ is the next one to be executed in some parallel component".
("λ is ready to execute".)

This informal explanation should already make clear that we are now able to express the information given by some program state $(\mu, \lambda_1, \ldots, \lambda_p, i)$, $i \neq 0$, by formulas of $\mathscr{L}_{TP\Pi}$, e.g., by

$$\text{at } \lambda_1 \wedge \ldots \wedge \text{at } \lambda_p \wedge \lambda_i,$$

(read "$\lambda_1, \ldots, \lambda_p$ are ready to execute and λ_i is executed next").

So we are able to describe a single state. As far as the second aspect noted at the beginning of this section is concerned, we will see shortly that the temporal operators of $\mathscr{L}_{TP\Pi}$ provide just what we wish to have.

According to the formal semantics of $\mathscr{L}_{TP\Pi}$ valid formulas of this language hold in any sequence of states. We are now interested in (additional) formulas holding for every "computation" of Π, i.e., in every temporal structure $\mathbf{K}=(\mathbf{S}, \xi, \mathbf{W})$ where $\mathbf{W}$ is just an execution sequence $\mathbf{W}_\Pi$ of Π w.r.t. $\mathbf{S}$ and ξ. Semantically we achieve this goal in the following way. Given some $\mathbf{S}$ and ξ, we interpret a program state

$$\eta_i = (\mu, \lambda_1, \ldots, \lambda_p, \kappa)$$

as a pair $\eta_i = (\eta_i^{(1)}, \eta_i^{(2)})$ of valuations as described in Section 11 with:

$\eta_i^{(1)}(a) = \mu(a)$ for program variables a,
$\eta_i^{(2)}(\lambda) = \mathbf{t}$ iff $\lambda = \lambda_\kappa$ (and $\kappa \neq 0$),
$\eta_i^{(2)}(\text{at } \lambda) = \mathbf{t}$ iff $\lambda = \lambda_j$ for some $j = 1, \ldots, p$.

Then we may apply our semantical apparatus and get, in fact, a formal notion of *validity in* $\mathbf{K} = (\mathbf{S}, \xi, \mathbf{W})$ where $\mathbf{W} = \mathbf{W}_\Pi$ is an execution sequence with respect to $\mathbf{S}$ and ξ.

However, this restriction to execution sequences or, equivalently, the validity of more formulas can also be described by giving additional axioms and rules which just allow for deriving these formulas. Actually we will give one rule and a set of axioms. For brevity, we call both axioms and rule *program axioms* and we may say that these program axioms *specify* the program Π.

We want to divide the program axioms into two classes:

- *structural axioms* (including the rule) which describe general properties of program states and execution sequences and hold for every program, and
- proper *specification axioms* which specify the execution sequences of some particular program.

In order to formulate rule and axioms we introduce the following abbreviations (assume $\bar{\mathcal{M}}_\Pi = \{\alpha_1, \ldots, \alpha_l\}$):

start_Π for $\text{at } \alpha_0^{(1)} \wedge \ldots \wedge \text{at } \alpha_0^{(p)} \wedge R$
("Execution of Π is in its initial state"),

nil_Π for $\neg\alpha_1 \wedge \neg\alpha_2 \wedge \ldots \wedge \neg\alpha_l$
("No action is executed").

Furthermore we assume $\mathcal{M}_{\Pi_i} = \{\alpha_0^{(i)}, \ldots, \alpha_{m_i}^{(i)}\}$ and let $\lambda, \lambda', \lambda_1, \lambda_2, \ldots, \text{at } \lambda$ be syntactic variables ranging over the set of propositional variables. Finally, we call any formula of the kernel $\mathcal{L}_\mathrm{P}$ of $\mathcal{L}_{\mathrm{TP}\Pi}$ (i.e., containing no temporal operators and no λ or at λ) a P-*formula* (using P, Q, R as syntactic variables for such formulas).

We now list the structural axioms describing the general properties of execution sequences.

(B1) $\text{start}_\Pi \to \Box A \vdash A$
(B2) $\text{nil}_\Pi \wedge A \to \circ(\text{nil}_\Pi \wedge A)$

(B1) is the rule we mentioned. It states that A is always true if it holds permanently beginning with the initial state of the execution sequence. This can be expressed as: "start_Π holds in the initial state (of $\mathbf{W}_\Pi$)". (B2) expresses that nothing is changed if no action is executed.

An obvious question is why we have given (B1) in the form of a rule and not as an axiom, say,

$$\Box(\text{start}_\Pi \to \Box A) \to A$$

so that the extension would be only by axioms, which is logically "more harmless" than adding rules. The reason is that, unfortunately, this latter formula turns out to be contradictory. Let $\mathbf{K}=(\mathbf{S}, \xi, \mathbf{W}_\Pi)$ be such that $\mathbf{K}_i(\text{start}_\Pi)=\mathbf{f}$ for $i \geq k$ with some $k \in \mathbb{N}_0$. Then $\mathbf{K}_k(\Box(\text{start}_\Pi \to \Box A))=\mathbf{t}$ for arbitrary A. The validity of $\Box(\text{start}_\Pi \to \Box A) \to A$ in $\mathbf{K}$ would imply that $\mathbf{K}_k(A)=\mathbf{t}$ for arbitrary A which can obviously not be the case. This fact has two immediate consequences:

- The Deduction Theorem no longer holds generally in the formal system $\Sigma_{\text{TP}}+(\text{B}1)$ resulting from Σ_{TP} by adding rule (B 1). (Otherwise the above formula would be derivable.)
- The effect of (B 1) cannot be axiomatized equivalently only by axioms. (Otherwise the Deduction Theorem would still hold, (B 1) and hence, again, the formula would be derivable.)

The first point prohibits free use of the Deduction Theorem in what we will call "program verification". In fact, we will never use it directly in our subsequent applications. It also implies that in the new context, logical laws which we have given or proved in the form $A \vdash B$ (e.g., (T 34), etc.) do not automatically imply that $\Box A \to B$ is also valid. However, it can be shown that for all our laws given in the previous chapters the latter form also remains valid, so we will continue to use these laws in both forms.

We call (B 1) and (B 2) *basic rule* and *basic axiom*, respectively, and separate them for a reason which will become clear later from the remaining structural axioms, which are:

(Π1) $\lambda \to \neg \lambda'$ for $\lambda \neq \lambda'$
("No two actions execute at the same time".)

(Π2) $\lambda \to \text{at } \lambda$
("An action may only execute if it is ready to".)

(Π3) $\text{at } \alpha_j^{(i)} \to \neg \text{at } \alpha_k^{(i)}$ for $j \neq k, j, k=0, \ldots, m_i, i=1, \ldots, p$
("In every Π_i, no two actions are ready to execute at the same time".)

(Π4) $\text{at } \lambda \wedge E_\lambda \to \neg \text{nil}_\Pi$
("If some action is *enabled* then some action has to execute".)

(Π5) $\text{at } \lambda \wedge \neg \lambda \to \circ \text{ at } \lambda$
("An action ready to execute but not executing remains ready".)

(Π6) $\text{at } \alpha_e^{(i)} \to \circ \text{ at } \alpha_e^{(i)}$ for $i=1, \ldots, p$
("If Π_i has reached its end label then it remains there".)

(Π7) $\lambda \wedge P \to \circ P$ if λ is the label of a statement $\notin \mathscr{A}$ and P is a P-formula
("Every statement $\notin \mathscr{A}$ does not cause any variable changes".)

The proper specification of a particular program Π consists of three parts:

- specification of the concrete possible sequences of locations (the "computation scheme" or "control flow" of Π),
- specification of the "data structure" $\mathbf{S}$ involved,
- specification of effects of the statements $\in \mathscr{A}$ of Π.

The computation scheme of Π is specified by the following axiom:

(CS) $\lambda \rightarrow (C_1 \wedge \circ \text{ at } \lambda_1) \vee \ldots \vee (C_q \wedge \circ \text{ at } \lambda_q)$
where $(\lambda, C_1, \lambda_1), \ldots, (\lambda, C_q, \lambda_q) \in \text{trans}(\Pi_i)$ for some $i = 1, \ldots, p$, and $\text{trans}(\Pi_i)$ contains no other element of the form $(\lambda, \ldots)$.

(Observe that (CS) appears more general than it really is: according to the syntax of our programs, q is always 1 or 2.) This axiom means: "If λ executes then execution proceeds one step according to the program structure given by $\text{trans}(\Pi_i)$" and describes the conditions and the location change of a single action.

So far, we have axiomatically mirrored what was meant by the semantic notion of validity in some $\mathbf{K} = (\mathbf{S}, \xi, \mathbf{W}_\Pi)$. Now we want to go a little bit further and assume that any execution sequence of Π presupposes some fixed interpretation of the "data" occurring in Π, i.e., the meaning of constants, function and predicate symbols and the range(s) of variables; formally we want a fixed structure $\mathbf{S}_\Pi$ for the underlying $\mathscr{L}_\mathrm{P}$. If, for example, Π works on natural numbers using a language with symbols $+$, $-$, $\times$, 0, 1, etc., then we suppose:

$|\mathbf{S}_\Pi| = \mathbb{N}_0$,
$\mathbf{S}_\Pi(+) =$ "plus" in $\mathbb{N}_0$,
etc.

Of course, there is no temporal aspect in this. The specification of $\mathbf{S}_\Pi$ is carried out by first-order formulas. However, it is known from classical logic that it might be impossible to really specify ("axiomatize") some particular $\mathbf{S}$ (e.g., the natural numbers) by finitely many axioms. Since these problems are not the subject of our present investigations we simply assume every formula which is valid in the particular $\mathbf{S}_\Pi$ to be a program axiom of Π. The set of these formulas is called the *theory of* $\mathbf{S}_\Pi$ and denoted by $\text{Th}(\mathbf{S}_\Pi)$. Thus we have as axioms:

(data) Every formula from $\text{Th}(\mathbf{S}_\Pi)$.

Furthermore, by our semantic definitions as well as by the program axioms, we have only specified program *schemes*, since we have not fixed the effect of a single action (except those $\notin \mathscr{A}$) on the variables, i.e., the change of the memory state μ in an execution sequence. If we fix some statement α to be, say, an assignment:

$\alpha \colon a := a + 1$

then, of course, every μ is changed to a well-defined μ' by executing α: $\mu'(a) = \mu(a) + 1$, $\mu'(b) = \mu(b)$ for $b \neq a$.

In the following we want to keep the freedom of fixing none, some, or all of the elementary statements of Π. An execution sequence where the relationships between μ and μ' according to all fixed statements are observed is called *legal*. (Thus, an execution sequence as defined in the previous section is a legal execution sequence for a program in which no elementary statement is fixed.)

The axiomatic specification of the effect of a concrete elementary statement on the memory state is in general given by (one or more) formulas of the form:

$\lambda \wedge P \rightarrow \circ Q \qquad (P, Q$ P-formulas).
("If λ executes in a state where P holds then Q holds after execution of λ".)

A standard situation will be that of an assignment:

$\lambda: a := t \qquad$ (a program variable, t term).

In this case we can describe the effect by the formula:

$$\lambda \wedge P_a(t) \rightarrow \circ P.$$

Example. Let $\lambda: a := 2 \times a$. Then we have, e.g., $\lambda \wedge b = 2 \times a \rightarrow \circ (b = a)$. □

The same holds, of course, if the assignment occurs within an **await** statement. Hence we have the following standard axioms:

(assign) $\lambda \wedge P_a(t) \rightarrow \circ P \qquad$ (P P-formula)
for every statement $\lambda: a := t$ or λ: **await** B **then** $a := t$ occurring in Π

We subsume under (assign) the case of:

λ: **await** B.

We may think of that as being the same as, say,

λ: **await** B **then** $a := a$

leading to:

$$\lambda \wedge P \rightarrow \circ P.$$

In non-standard situations where an elementary statement is of another type we will specify it by one or more other particular axioms.

Let us summarize now all these discussions. Given a program Π (possibly with elementary statements left open, denoted by $\lambda: \sim$) we call a formula A of $\mathscr{L}_{\mathrm{TP}\Pi}$ *Π-valid* if A is valid in every temporal structure $\mathbf{K} = (\mathbf{S}, \xi, \mathbf{W})$, where

- $\mathbf{S}$ is the structure $\mathbf{S}_\Pi$ determined by Π,
- $\mathbf{W}$ is a legal execution sequence of Π.

The generalization to the definition of *A follows in Π from* a set $\mathscr{F}$ of formulas (briefly denoted by $\mathscr{F} \Vdash_\Pi A$) is as usual. Furthermore, we let Spec(Π) be the set of program axioms (except rule (B1)) corresponding to Π and call *A Π-derivable* (briefly $\Pi \vdash A$) if

$$\mathrm{Spec}(\Pi) \vdash A$$

in the formal system $\Sigma_{\mathrm{TP}} + (\mathrm{B}1)$. Spec($\Pi$) together with (B1) is also called the *temporal semantics* of Π. The soundness of the notion of Π-derivability with respect to Π-validity is shown by the following theorem.

Theorem 14.1. (Soundness Theorem for Spec(Π) and (B1).) *Let Π be a program, A a formula of $\mathscr{L}_{\mathrm{TP}\Pi}$ and $\mathscr{F}$ a set of formulas of $\mathscr{L}_{\mathrm{TP}\Pi}$. If $\mathscr{F} \cup \mathrm{Spec}(\Pi) \vdash A$ in $\Sigma_{\mathrm{TP}} + (\mathrm{B}1)$ then $\mathscr{F} \Vdash_\Pi A$.*

Proof. Analogously to previous soundness proofs we show that all axioms of Spec(Π) are Π-valid and that (B1) preserves validity. Let $\mathbf{K}=(\mathbf{S}_\Pi, \xi, \mathbf{W}_\Pi)$ be a temporal structure with legal $\mathbf{W}_\Pi=\{\eta_0, \eta_1, \eta_2, \dots\}$.

(B1): $\Vdash_{\mathbf{K}} \text{start}_\Pi \to \Box A \Rightarrow \mathbf{K}_0(\Box A)=\mathbf{t}$ since $\mathbf{K}_0(\text{start}_\Pi)=\mathbf{t}$
$\Rightarrow \mathbf{K}_i(A)=\mathbf{t}$ for all $i\in\mathbb{N}_0$
$\Rightarrow \Vdash_{\mathbf{K}} A.$

(B2): $\mathbf{K}_i(\text{nil}_\Pi \wedge A)=\mathbf{t} \Rightarrow \eta_i=(\mu, \lambda_1, \dots, \lambda_p, 0)$
$\Rightarrow \eta_{i+1}=\eta_i$
$\Rightarrow \mathbf{K}_{i+1}(\text{nil}_\Pi \wedge A)=\mathbf{t}$
$\Rightarrow \mathbf{K}_i(\circ(\text{nil}_\Pi \wedge A))=\mathbf{t}.$

(Π1): $\mathbf{K}_i(\lambda)=\mathbf{t} \Rightarrow \lambda=\lambda_\kappa$ in η_i
$\Rightarrow \lambda' \neq \lambda_\kappa$ in η_i
$\Rightarrow \mathbf{K}_i(\lambda')=\mathbf{f}$
$\Rightarrow \mathbf{K}_i(\neg\lambda')=\mathbf{t}.$

(Π2): $\mathbf{K}_i(\lambda)=\mathbf{t} \Rightarrow \lambda=\lambda_\kappa$ in η_i
$\Rightarrow \mathbf{K}_i(\text{at }\lambda)=\mathbf{t}.$

(Π3): $\mathbf{K}_i(\text{at }\alpha_j^{(l)})=\mathbf{t} \Rightarrow \alpha_j^{(l)}=\lambda_l$ in η_i
$\Rightarrow \alpha_k^{(l)} \neq \lambda_l$ in η_i
$\Rightarrow \mathbf{K}_i(\text{at }\alpha_k^{(l)})=\mathbf{f}$
$\Rightarrow \mathbf{K}_i(\neg \text{ at }\alpha_k^{(l)})=\mathbf{t}.$

(Π4): $\mathbf{K}_i(\text{at }\lambda \wedge E_\lambda)=\mathbf{t} \Rightarrow \lambda=\lambda_j$ for some j and $\mathbf{S}^{(\xi,\mu)}(E_\lambda)=\mathbf{t}$ in η_i
$\Rightarrow \lambda_j \neq \alpha_e^{(j)}$ and $\mathbf{S}^{(\xi,\mu)}(E_\lambda)=\mathbf{t}$ in η_i
$\Rightarrow \kappa \neq 0$ in η_i
$\Rightarrow \mathbf{K}_i(\lambda_\kappa)=\mathbf{t}$
$\Rightarrow \mathbf{K}_i(\text{nil}_\Pi)=\mathbf{f}$
$\Rightarrow \mathbf{K}_i(\neg\text{nil}_\Pi)=\mathbf{t}.$

(Π5): $\mathbf{K}_i(\text{at }\lambda \wedge \neg\lambda)=\mathbf{t} \Rightarrow \lambda=\lambda_j$ for some j and $\lambda \neq \lambda_\kappa$ in η_i
$\Rightarrow \lambda=\lambda_j$ for the same λ_j in η_{i+1}
$\Rightarrow \mathbf{K}_{i+1}(\text{at }\lambda)=\mathbf{t}$
$\Rightarrow \mathbf{K}_i(\circ \text{ at }\lambda)=\mathbf{t}.$

(Π6): $\mathbf{K}_i(\text{at }\alpha_e^{(l)})=\mathbf{t} \Rightarrow \lambda_l=\alpha_e^{(l)}$ in η_i
$\Rightarrow \lambda_l=\alpha_e^{(l)}$ and $\kappa \neq l$ in η_i since trans(Π_l) contains no element of the form $(\alpha_e^{(l)}, \dots)$
$\Rightarrow \lambda_l=\alpha_e^{(l)}$ in η_{i+1}
$\Rightarrow \mathbf{K}_i(\circ \text{ at }\alpha_e^{(l)})=\mathbf{t}.$

(Π7): $\mathbf{K}_i(\lambda \wedge P)=\mathbf{t} \Rightarrow \lambda=\lambda_\kappa$ in $\eta_i=(\mu, \dots)$ and $\mathbf{K}_i(P)=\mathbf{S}^{(\xi,\mu)}(P)=\mathbf{t}$
$\Rightarrow \mathbf{K}_{i+1}(P)=\mathbf{S}^{(\xi,\mu')}(P)=\mathbf{t}$ for $\eta_{i+1}=(\mu', \dots)$ since $\mu=\mu'$
$\Rightarrow \mathbf{K}_i(\circ P)=\mathbf{t}.$

(CS): $\mathbf{K}_i(\lambda)=\mathbf{t} \Rightarrow \eta_i=(\mu, \dots, \lambda, \dots, \kappa)$, $\kappa \neq 0$
$\Rightarrow \eta_{i+1}=(\mu', \dots, \lambda_r, \dots, \kappa')$, $(\lambda, C_r, \lambda_r)\in\text{trans}(\Pi_\kappa)$ and $\mathbf{S}^{(\xi,\mu)}(C_r)=\mathbf{t}$ for some $r=1, \dots, q$
$\Rightarrow \mathbf{K}_i(C_r)=\mathbf{t}$ and $\mathbf{K}_{i+1}(\text{at }\lambda_r)=\mathbf{t}$ for some $r=1, \dots, q$
$\Rightarrow \mathbf{K}_i((C_1 \wedge \circ \text{ at }\lambda_1) \vee \dots \vee (C_q \wedge \circ \text{ at }\lambda_q))=\mathbf{t}.$

The axioms under (data) are trivially Π-valid, and the specification axioms for elementary statements can be viewed to be Π-valid by definition. In the standard case of an assignment $a:=t$, the axiom (assign) meets the informal meaning:

$$\begin{aligned}\mathbf{K}_i(\lambda \wedge P_a(t))=\mathbf{t} &\Rightarrow \eta_i=(\mu, \ldots), \quad \mathbf{S}^{(\xi,\mu)}(P_a(t))=\mathbf{t} \text{ and } \eta_{i+1}=(\mu', \ldots)\\ &\quad \text{with } \mu'(a)=\mathbf{S}^{(\xi,\mu)}(t) \text{ and } \mu'(b)=\mu(b) \quad \text{for } b \neq a\\ &\Rightarrow \mathbf{S}^{(\xi,\mu')}(P)=\mathbf{S}^{(\xi,\mu)}(P_a(t))=\mathbf{t}\\ &\Rightarrow \mathbf{K}_i(\circ P)=\mathbf{t}. \quad \square\end{aligned}$$

When subsequently proving the Π-derivability of formulas we will, in order to facilitate derivations, feel free to condense trivial steps to one.

Example. Let α: $a:=2\times a$. We will not hesitate to state that, say,

$$\alpha \wedge b=0 \wedge a=k \rightarrow \circ(b=0 \wedge a=2\times k)$$

is derivable. This fact is intuitively clear and could be formally shown as follows:

(1)	$\alpha \wedge b=0 \wedge 2\times a=2\times k \rightarrow \circ(b=0 \wedge a=2\times k)$	(assign)
(2)	$a=k \rightarrow 2\times a=2\times k$	(data)
(3)	$\alpha \wedge b=0 \wedge a=k \rightarrow \circ(b=0 \wedge a=2\times k)$	(prop), (1), (2) $\square$

Other typical cases are assertions about the control flow in Π.

Example. Let α_1 : **if** $a \neq 0$ **then** α_2: $\sim$ **else** α_3 : $\sim$ **fi**. We might state the following "rule":

at $\alpha_1 \rightarrow a>0 \vdash \alpha_1 \rightarrow \circ$ at α_2.

("Infer that execution of α_1 leads to α_2 if always $a>0$ at location α_1".)

A complete derivation could look like this:

(1)	at $\alpha_1 \rightarrow a>0$	assumption
(2)	$\alpha_1 \rightarrow (a\neq 0 \wedge \circ$ at $\alpha_2) \vee (a=0 \wedge \circ$ at $\alpha_3)$	(CS)
(3)	$\alpha_1 \rightarrow$ at α_1	(Π2)
(4)	$\alpha_1 \rightarrow a\neq 0$	(data), (prop), (1), (3)
(5)	$\alpha_1 \rightarrow \circ$ at α_2	(prop), (2), (4) $\square$

In any case we will indicate the use of such formulas or conclusions which can "easily be seen from the program text" by (Π). Moreover, an important tool will be further derived proof rules:

$$A_1, \ldots, A_n \vdash B$$

which, however, are not derivable in Σ_{TP} itself but only in the context of the additional program axioms of some Π, i.e., stemming from the derivability of B from $\mathrm{Spec}(\Pi) \cup \{A_1, \ldots, A_n\}$ in Σ_{TP}+(B1). We call such rules *Π-derived* and may use them in derivations of Π-derivable formulas.

We conclude this section by stating a very useful rule of this kind:

(trans) $\alpha \wedge A \rightarrow \circ B$ for every $\alpha \in \bar{\mathcal{M}}_{\Pi}$,
$\mathrm{nil}_{\Pi} \wedge A \rightarrow B$
$\vdash A \rightarrow \circ B$.

Derivation of (trans)
(Suppose $\bar{\mathcal{M}}_{\Pi} = \{\alpha_1, \ldots, \alpha_n\}$.)

(1)	$\alpha \wedge A \rightarrow \circ B$ for every $\alpha \in \bar{\mathcal{M}}_{\Pi}$	assumption
(2)	$\mathrm{nil}_{\Pi} \wedge A \rightarrow B$	assumption
(3)	$\mathrm{nil}_{\Pi} \vee \alpha_1 \vee \ldots \vee \alpha_n$	(taut)
(4)	$\mathrm{nil}_{\Pi} \wedge A \rightarrow \mathrm{nil}_{\Pi} \wedge B$	(prop), (2)
(5)	$\mathrm{nil}_{\Pi} \wedge B \rightarrow \circ B$	(prop), (B2)
(6)	$\mathrm{nil}_{\Pi} \wedge A \rightarrow \circ B$	(prop), (4), (5)
(7)	$A \rightarrow \circ B$	(prop), (1), (3), (6) □

It should be noticed that in the derivation of (trans) the only program axiom which is needed is the basic axiom (B2).

15. Description of Program Properties

Formulas of the language $\mathcal{L}_{\mathrm{TP}\Pi}$ express properties of execution sequences of a program Π, or briefly, properties of Π. If such a formula A is Π-derivable then we may say that Π has the property described by A. In the subsequent chapters we want to investigate how properties of programs can be proved. In this section we want to give a first cursory overview of what properties might be of interest.

A first fundamental classification of program properties is as follows:

- *Safety* (or *invariance*) *properties*,
- *Liveness* (or *eventuality*) *properties*.

We give some illustrations and examples for these classes.

(a) Safety properties

These properties are expressed by formulas of the form:

$$A \rightarrow \Box B.$$

A special case is given if $A \equiv \mathbf{true}$, hence, the formula is reduced to

$$\Box B.$$

But because of $\Box B \vdash \mathrm{start}_{\Pi} \rightarrow \Box B$ and also $\mathrm{start}_{\Pi} \rightarrow \Box B \vdash \Box B$ (with (B1) and (alw)), $\Box B$ and $\mathrm{start}_{\Pi} \rightarrow \Box B$ are equivalent w.r.t. derivability. So we will often use $\mathrm{start}_{\Pi} \rightarrow \Box B$ instead of $\Box B$ since it has again the standard form and is also somewhat more intuitive ("B holds permanently from the program start").

(a1) Partial correctness

Let Π be a non-cyclic program, P and Q P-formulas. We want to express the following fact:

> "If P holds upon the start of a computation of Π
> (additionally to the initialization condition R of Π)
> and the computation *terminates* (i.e., reaches the end labels)
> then Q holds upon termination".

This property is called *partial correctness of Π w.r.t. the precondition P and the postcondition Q*. It can be expressed in $\mathscr{L}_{\mathrm{TP}\Pi}$ by the formula:

$$\mathrm{start}_\Pi \wedge P \to \Box(\mathrm{at}\ \alpha_e^{(1)} \wedge \ldots \wedge \mathrm{at}\ \alpha_e^{(p)} \to Q)$$

("Starting with P, Q will hold whenever execution is at the end labels".)

Example. Let

$\Pi \equiv$ **initial** $a=n \wedge b=m$;
⋮
α_e: **stop**

be a sequential program computing on variables q and r, the quotient and remainder of n/m for input values n, $m \in \mathbb{N}_0$, $m>0$. The corresponding assertion of partial correctness is:

$$\mathrm{start}_\Pi \wedge m>0 \to \Box(\mathrm{at}\ \alpha_e \to n=q\times m+r \wedge 0\leq r<m). \quad \Box$$

(a2) Global and generalized invariants

The formula of partial correctness above expresses the fact that some Q holds in the terminal state of any computation of Π. This can be modified in two directions:

- Q is to be true in every state,
- Q is to be true in certain states which are to be described.

The first assertion can be formulated in $\mathscr{L}_{\mathrm{TP}\Pi}$ as:

$$\mathrm{start}_\Pi \wedge P \to \Box Q$$

and we call Q in this case a *global invariant* (*w.r.t. precondition P*). An example for the second kind of assertions could be:

$$\mathrm{start}_\Pi \wedge P \to \Box(\mathrm{at}\ \alpha_1 \vee \ldots \vee \mathrm{at}\ \alpha_m \to Q).$$

("Q holds at every location $\alpha_1, \ldots, \alpha_m$".)

Other possibilities are evident. In every comparable case we call Q a *generalized invariant.*

Example. Let α_1 and α_2 be labels of statements in some Π, which perform a division by some variable a. The property that this division is always possible is expressed by

$$\mathrm{start}_\Pi \wedge P \to \Box(\mathrm{at}\ \alpha_1 \vee \mathrm{at}\ \alpha_2 \to a \neq 0). \quad \Box$$

(a3) Mutual exclusion

We discuss this property by a sample situation. Let Π be a program with two parallel components Π_1, Π_2 of the form:

$$\begin{array}{ll} \Pi_1 \equiv \alpha_0 : \sim; & \Pi_2 \equiv \beta_0 : \sim; \\ \quad\vdots & \quad\vdots \\ \quad\alpha_i : \sim; & \quad\beta_k : \sim; \\ \quad\vdots & \quad\vdots \\ \quad\alpha_j : \sim; & \quad\beta_l : \sim; \\ \quad\vdots & \quad\vdots \end{array}$$

Suppose the sections $\alpha_i - \alpha_j$ and $\beta_k - \beta_l$ are *critical*, meaning that Π_1 and Π_2 must not be in these sections at the same time. This *mutual exclusion* is expressed by:

$$\text{start}_\Pi \wedge P \rightarrow \Box\neg((\text{at } \alpha_i \vee \ldots \vee \text{at } \alpha_j) \wedge (\text{at } \beta_k \vee \ldots \vee \text{at } \beta_l)).$$

(Here and in the following P is, as before, some additional, possibly "empty" precondition.)

We may interpret mutual exclusion in another way. The blocks of statements $\alpha_i - \alpha_j$ and $\beta_k - \beta_l$ are "indivisible" with respect to each other (remember our discussion of the interleaving model in Section 12). How this can be really achieved will be seen in the next section.

(a4) Deadlock freedom

Again we give a simple example. Let Π consist of two cyclic parallel components of the form:

$$\begin{array}{ll} \Pi_1 \equiv & \Pi_2 \equiv \\ \quad\vdots & \quad\vdots \\ \quad\alpha_i : \texttt{await } B_1; & \quad\beta_k : \texttt{await } B_2; \\ \quad\vdots & \quad\vdots \end{array}$$

A deadlock of Π occurs if Π_1 and Π_2 are at location α_i and β_k, respectively, and both B_1 and B_2 are false. The property that this will not happen (*deadlock freedom*) is expressed by:

$$\text{start}_\Pi \wedge P \rightarrow \Box(\text{at } \alpha_i \wedge \text{at } \beta_k \rightarrow B_1 \vee B_2).$$

(b) Liveness properties

These properties are expressed by formulas of the form:

$$A \rightarrow \Diamond B$$

stating the *existence* of some state in which B holds.

(b1) Total correctness and termination

For a non-cyclic program Π and P-formulas P and Q the *total correctness of Π w.r.t. precondition P and postcondition Q* is the property:

> "If P holds upon the start of a computation of Π then the computation terminates and Q holds upon termination".

(Observe the difference between partial and total correctness.) This property is expressed by:

$$\text{start}_\Pi \wedge P \rightarrow \Diamond(\text{at } \alpha_e^{(1)} \wedge \ldots \wedge \text{at } \alpha_e^{(p)} \wedge Q).$$

Example. In our example illustrating partial correctness above we could state the total correctness by:

$$\text{start}_\Pi \wedge m > 0 \rightarrow \Diamond(\text{at } \alpha_e \wedge n = q \times m + r \wedge 0 \leq r < m). \quad \square$$

A "subnotion" of total correctness is the property of *termination* alone:

$$\text{start}_\Pi \wedge P \rightarrow \Diamond(\text{at } \alpha_e^{(1)} \wedge \ldots \wedge \text{at } \alpha_e^{(p)}).$$

(b2) More general accessibility properties

The properties under (b1) can be generalized in manifold ways. We note only two special cases; others will occur later. The first case is:

$\text{at } \alpha \rightarrow \Diamond \text{ at } \alpha'$.
("If execution is at α then it will sometime reach α'".)

We call this *accessibility* of α' from α. The second case is:

$\text{at } \alpha \wedge P \rightarrow \Diamond(\text{at } \alpha' \wedge Q)$.
("If execution is at α and P holds then it will sometime reach α' with Q being true".)

P and Q are sometimes called *intermittent assertions* in comparable cases (but see also Section 25 for a special meaning of this notion).

Example. Let Π be the program illustrating mutual exclusion above. Let α_{i-1} be the label immediately preceeding α_i. A trivial (but uninteresting) means of achieving mutual exclusion is to take:

α_{i-1}: `await false`.

Then Π_1 will never enter its critical region. If, however, we want to demand that Π_1, "waiting" at α_{i-1}, will eventually enter the section $\alpha_i - \alpha_j$, we can express this property by:

$$\text{at } \alpha_{i-1} \rightarrow \Diamond \text{ at } \alpha_i. \quad \square$$

According to their syntactic structure, safety and liveness properties are in some sense "dual" to each other. This is seen even better if we write a safety property $A \rightarrow \Box B$ in the form $A \rightarrow \neg \Diamond \neg B$ or:

$$A \rightarrow \neg \Diamond B'.$$

This expresses that "something (bad) will never happen", whereas a liveness property:

$$A \rightarrow \Diamond B$$

states that "something (good) will eventually happen".

Of course, there are interesting properties of programs other than safety and liveness. A third class which has received attention is the class of *precedence properties* describing some precedence relationship between the occurrence of certain events and expressed by use of the temporal operators **atnext**, **unless**, etc. Again, we give some typical examples.

(c) Precedence properties

Analogously to safety and liveness formulas the simplest form of precedence properties is:

$$A \rightarrow B \ \mathbf{atnext}\ C$$

or, using other operators,

$$A \rightarrow B \ \mathbf{unless}\ C, \text{ etc.}$$

(c1) Sequences of assertions

Consider once more a general invariance property of the kind:

$$\text{start}_{\Pi} \wedge P \rightarrow \Box(A \rightarrow Q).$$

("Every time when A holds then Q holds".)

More generally we may think of a situation where not necessarily the *same* Q holds at all points with A holding but some sequence $Q_0, Q_1, Q_2, \ldots$ of assertions holds instead:

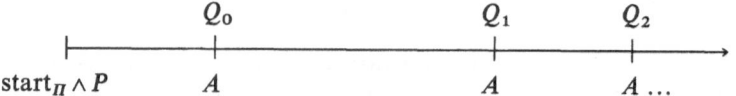

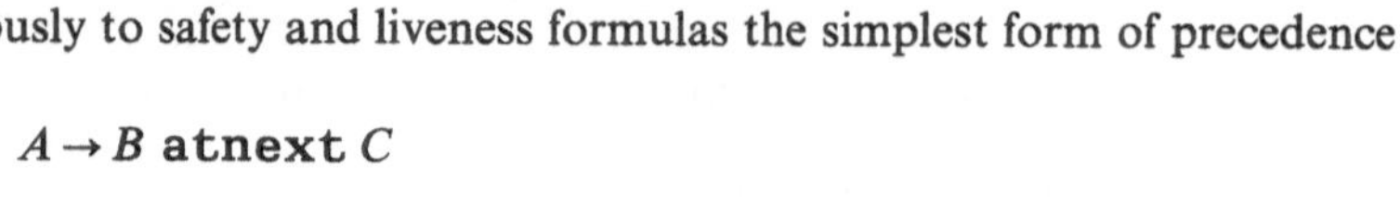

The property is expressed by the two formulas:

$$\text{start}_{\Pi} \wedge P \rightarrow Q_0 \ \mathbf{atnext}\ A,$$
$$A \wedge Q_i \rightarrow Q_{i+1} \ \mathbf{atnext}\ A.$$

Example. Let

```
Π ≡ initial k=0;
    loop
        ⋮
        α1: if "good" then α2: output(k) fi;
        α3: k:=k+1
    end
```

be a sequential program testing the sequence of natural numbers with respect to some desired property good(n) and outputting the sequence of "good" ones. Defining the function:

$$\text{nextgood}: \mathbb{N}_0 \rightarrow \mathbb{N}_0$$

by nextgood(n) = smallest $m \in \mathbb{N}_0$ with $m \geq n$ and good(m), we can express the intended effect of Π by:

$$\mathrm{start}_\Pi \rightarrow k = \mathrm{nextgood}(0) \;\mathbf{atnext}\; \mathrm{at}\ \alpha_2,$$
$$\mathrm{at}\ \alpha_2 \wedge k = k_0 \rightarrow k = \mathrm{nextgood}(k_0 + 1) \;\mathbf{atnext}\; \mathrm{at}\ \alpha_2. \quad \square$$

(c2) More general precedence relationships

We discuss some possible situations using an example. Let Π be a program with parallel components $\Pi_1, \ldots, \Pi_q, \Psi$. Think of every Π_i, $i = 1, \ldots, q$, as a process requesting from time to time some resource, and of Ψ being the "granter" of this resource which exists in a finite number of equivalent instances, i.e., each Π_i is of the form:

```
Πᵢ ≡ loop
          ⋮
     α₁⁽ⁱ⁾: "request resource";
          ⋮
     α₂⁽ⁱ⁾: "get resource";
          ⋮
     α₃⁽ⁱ⁾: "release resource";
          ⋮
     end
```

and Ψ is of the form:

```
Ψ ≡ loop
         ⋮
     β: grant(k)
         ⋮
    end
```

where the statement grant(k) grants the resource to Π_k.

Of course, there should be some conditions upon a reasonable cooperation of these processes. Let L_i be the set of labels between $\alpha_1^{(i)}$ (not included) and $\alpha_2^{(i)}$ (included). If $L_i = \{\gamma_1, \ldots, \gamma_l\}$ we let

$$\mathrm{at}\ L_i \equiv \mathrm{at}\ \gamma_1 \vee \ldots \vee \mathrm{at}\ \gamma_l.$$

Informally, at L_i means: "Π_i has requested the resource and is waiting to get it".

A first simple condition would be that Ψ grants the resource only to a process which has really requested it. This can be expressed by:

$$\neg\, \mathrm{at}\ L_i \rightarrow \neg (\beta \wedge k = i) \wedge \alpha_1^{(i)} \;\mathbf{before}\; (\beta \wedge k = i) \quad \text{(for every } i = 1, \ldots, q).$$

("If Π_i is not waiting for a grant then the resource is not granted to Π_i now, and there must be a request from Π_i before a later grant to Π_i".)

At first glance, one might think of a simpler description of this property by the formula:

$$\beta \wedge k = i \rightarrow \mathrm{at}\ L_i.$$

("If Ψ grants the resource to Π_i then Π_i is waiting for it".)

However, this formula does not meet the intention precisely: Ψ could grant a second instance of the resource to Π_i before Π_i has used the first one.

Of particular interest is the "fair" behaviour of the granter Ψ with respect to the order of arriving requests. One possible strategy is that of first-come-first-served, i.e., if the request of Π_i occurs before the request of Π_j then Π_i is served first. This can be expressed by the formula:

$\alpha_1^{(i)} \wedge \neg$ at $L_j \rightarrow \neg(\beta \wedge k=j)$ **unless** $(\beta \wedge k=i)$.
("If Π_i requests and Π_j is not waiting for the resource then it is not granted to Π_j until a grant is made to Π_i".)

It is remarkable that such precedence properties can mostly be described by different operators in different ways. For example, other formulations of the latter property could be:

$\alpha_1^{(i)} \wedge \neg$ at $L_j \rightarrow (\beta \wedge k=i)$ **before** $(\beta \wedge k=j)$,
("If Π_i requests and Π_j is not waiting for the resource then it will be granted to Π_i before a grant to Π_j".)

$\alpha_1^{(i)} \wedge \neg$ at $L_j \rightarrow \neg(\beta \wedge k=j)$ **while** $\neg(\beta \wedge k=i)$,
("If Π_i requests and Π_j is not waiting for the resource then it is not granted to Π_j as long as it is not granted to Π_i".)

$\alpha_1^{(i)} \wedge \neg$ at $L_j \rightarrow k=i$ **atnext** $(\beta \wedge (k=i \vee k=j))$.
("If Π_i requests and Π_j is not waiting for the resource then the next grant to Π_i or Π_j goes to Π_i".)

Moreover, properties of this kind may be differently expressed even by the same operator, e.g., in our case by:

$\alpha_1^{(i)}$ **before** at $L_j \rightarrow (\beta \wedge k=i)$ **before** $(\beta \wedge k=j)$,

the informal interpretation of which is left to the reader.

A somewhat more liberal strategy than first-come-first-served (which might be difficult to implement) would allow some maximal number of "overtakings". For example, we could allow Π_j to overtake Π_i, but once at most (*1-bounded overtaking*). This can be expressed by the formula:

$\alpha_1^{(i)} \wedge \neg$at $L_j \rightarrow k=i$ **atnext** $(\beta \wedge (k=i \vee k=j)) \vee$
$k=i$ **atnext**^{2} $(\beta \wedge (k=i \vee k=j))$.
("If Π_i requests and Π_j is not waiting for the resource then the next or the second next grant to Π_i or Π_j goes to Π_i".)

Observe that this formula is not strictly of the form $A \rightarrow B$ **atnext** C but a slight generalization of it. Again, there are other possible descriptions of the same property, e.g., by the nested unless formula:

$\alpha_1^{(i)} \wedge \neg$ at $L_j \rightarrow \neg(\beta \wedge k=j)$ **unless** $[(\beta \wedge k=j)$ **unless**
$[\neg(\beta \wedge k=j)$ **unless** $(\beta \wedge k=i)]]$.

Quite analogously one can express *n-bounded overtaking* for $n>1$.

We want to conclude with a short summary of this chapter:

- The *language* $\mathscr{L}_{TP\Pi}$ enables us to *describe* properties of programs.
- The *logic* $\mathscr{L}_{TP\Pi}$ (i.e., the formal system Σ_{TP} plus the additional program specification rule and axioms) enables us to *prove* that a program really has some property.

The remaining part of this book is dedicated to the latter aspect: methods and examples of *program verification.* As already indicated in Section 7, invariance and precedence properties (because of their "weak" character) are similar in their linguistic intention and also involve similar proof methods. In fact, both should be viewed as (a wider class of) safety properties in the sense that they express "if something happens at all it happens correctly". We will treat them together in the following chapter and separate them from the essentially different class of liveness properties investigated in Chapter VI.

Chapter V

Invariance and Precedence Properties of Programs

16. The Basic Invariant Method

Let

$$\Pi \equiv \texttt{initial}\ R;\ \texttt{cobegin}\ \Pi_1 \parallel \dots \parallel \Pi_p\ \texttt{coend}$$

be a program. We begin with the consideration of invariance properties whose general form is

$$A \rightarrow \Box B.$$

In Section 7 we already noted a general logical proof principle for proving such formulas:

(ind″) $A \rightarrow B,\ B \rightarrow \circ B \vdash A \rightarrow \Box B.$

We have only to make precise how $B \rightarrow \circ B$ can be established in the context of a program. For this purpose we remember the rule:

(trans) $\alpha \wedge A \rightarrow \circ B$ for every $\alpha \in \bar{\mathscr{M}}_\Pi$,
$\mathrm{nil}_\Pi \wedge A \rightarrow B$
$\vdash A \rightarrow \circ B$

from Section 14 which solves just this task.

We introduce the following abbreviations:

$C\ \texttt{invof}\ \alpha$ for $\alpha \wedge C \rightarrow \circ C$
("C is an *invariant of* α"),

$C\ \texttt{invof}\ \mathscr{M}$ for $C\ \texttt{invof}\ \alpha_1 \wedge \dots \wedge C\ \texttt{invof}\ \alpha_n$
where $\mathscr{M} = \{\alpha_1, \dots, \alpha_n\}$
("C is an invariant of every $\alpha \in \mathscr{M}$"),

and combine (ind″) and (trans) to an *invariant rule*:

(inv) $A \rightarrow B$,
$B\ \texttt{invof}\ \bar{\mathscr{M}}_\Pi$
$\vdash A \rightarrow \Box B.$

This rule expresses a simple and basic invariant method: "In order to prove that B holds in every execution state from a state in which A holds, show that B is true in this first state and is invariant under every action". The validity of (inv) is intuitively clear, but we also want to state and prove this formally.

Theorem 16.1. *The rule (inv) is Π-derived.*

Proof. We give a direct derivation:

(1)	$A \rightarrow B$		assumption
(2)	B **invof** $\bar{\mathscr{M}}_\Pi$		assumption
(3)	$\alpha \wedge B \rightarrow \circ B$	for every $\alpha \in \bar{\mathscr{M}}_\Pi$	(prop), (2)
(4)	$\mathrm{nil}_\Pi \wedge B \rightarrow B$		(taut)
(5)	$B \rightarrow \circ B$		(trans), (3), (4)
(6)	$A \rightarrow \Box B$		(ind″), (1), (5) □

All examples of invariance properties in the previous section were actually of the form:

$$\mathrm{start}_\Pi \wedge P \rightarrow \Box A.$$

For these formulas (inv) reads:

$$\mathrm{start}_\Pi \wedge P \rightarrow A,$$
$$A \ \mathbf{invof}\ \bar{\mathscr{M}}_\Pi$$
$$\vdash \mathrm{start}_\Pi \wedge P \rightarrow \Box A.$$

In the case that P is empty (i.e., **true**) we may either derive a formula $\mathrm{start}_\Pi \rightarrow \Box A$ or continue the derivation a little bit and get the formula $\Box A$. This latter aspect is formulated by:

(inv′) $\mathrm{start}_\Pi \rightarrow A$, A **invof** $\bar{\mathscr{M}}_\Pi \vdash \Box A$.

Derivation of (inv′). By (inv) we get from the two premises:

(1)	$\mathrm{start}_\Pi \rightarrow \Box A$	
(2)	A	(B1), (1)
(3)	$\Box A$	(alw), (2) □

This derivation is so trivial, that we will also feel free to apply it implicitly; having derived some $\mathrm{start}_\Pi \rightarrow \Box A$, we will use $\Box A$ when this is more convenient.

An interesting observation is that these proof rules are based on a purely logical induction principle and minimal information about the execution sequences of the particular program. In fact, deriving (inv) and (inv′) we used only the two basic (B1) and (B2), the latter not explicitly but via (trans).

Now let us illustrate actual use of the invariant method with a first simple example. We choose the mutual exclusion problem already mentioned in previous sections. The pattern of the problem is that the components Π_1 and Π_2 of some program Π are cyclic loops, the bodies of which contain critical sections cs_1 and cs_2, respectively, which are to be mutually excluded in computations of Π. (The generalization to the same situation with more parallel components is obvious.) One possible basic scheme of solving this problem is given by the following program:

```
Π ≡ initial ex=true;
    cobegin loop α0: ~;
                       ⋮
                 α1: await ex=true then ex:=false;
                       ⋮}cs1
                 α2: ex:=true;
                       ⋮
            end
           ||
            loop β0: ~;
                       ⋮
                 β1: await ex=true then ex:=false;
                       ⋮}cs2
                 β2: ex:=true;
                       ⋮
            end
    coend
```

The actions α_1, α_2, β_1, β_2 manage the desired synchronization. We may assume that they are the only actions in which the variable *ex* occurs.

Let us briefly come back to our discussion of the interleaving model in Section 12. Whenever in subsequent program examples we have "complicated" elementary statements and whenever their indivisibility (contradicting intuition) is really used, we may assume that these statements stand for greater blocks of more elementary statements "protected" (i.e., made indivisible) against other "critical" statements by a synchronization like the above. We have only to be sure that there are "special" variables like *ex* on which the operations:

await *ex*=**true** **then** *ex*:=**false**, and
ex:=**true**

are in fact indivisible. Such variables are called *binary semaphores*, and the operations are usually denoted by P(*ex*) and V(*ex*), respectively.

We now want to prove formally that Π has indeed the desired property. Let us first introduce some more abbreviations. If $\mathcal{M}=\{\alpha_1, \ldots, \alpha_n\}$ is a set of labels, we write:

at $\mathcal{M}$ for at $\alpha_1 \vee \ldots \vee$ at α_n.
("One of the $\alpha \in \mathcal{M}$ is ready to execute".)

Furthermore we introduce the following (classical) propositional operators:

exor $(A_1, \ldots, A_k)$ for $(A_1 \wedge \neg A_2 \wedge \ldots \wedge \neg A_k) \vee$
$(\neg A_1 \wedge A_2 \wedge \ldots \wedge \neg A_k) \vee$
⋮
$(\neg A_1 \wedge \ldots \wedge \neg A_{k-1} \wedge A_k)$
("Exactly one of $A_1, \ldots, A_k$ is true"),

excl $(A_1, \ldots, A_k)$ for **exor** $(A_1, \ldots, A_k) \vee (\neg A_1 \wedge \ldots \wedge \neg A_k)$
("At most one of $A_1, \ldots, A_k$ is true").

It is an easy exercise in propositional logic to check the following tautologies:

$$\mathbf{excl}\,(A_1, \ldots, A_k) \to \mathbf{excl}\,(A_1, \ldots, A_{k-1}),$$
$$\mathbf{excl}\,(B_1 \vee B_2, A_1, \ldots, A_k) \to \mathbf{excl}\,(B_1, A_1, \ldots, A_k).$$

Now let L_1 and L_2 be the sets of labels in the critical sections cs_1 and cs_2 of Π, respectively. The assertion of mutual exclusion is:

$$\Pi \vdash \mathrm{start}_\Pi \to \Box\,\mathbf{excl}\,(\mathrm{at}\ L_1, \mathrm{at}\ L_2).$$

(In Section 15 we had $\mathrm{start}_\Pi \to \Box\neg(\mathrm{at}\ L_1 \wedge \mathrm{at}\ L_2)$. Obviously $\neg(\mathrm{at}\ L_1 \wedge \mathrm{at}\ L_2)$ is equivalent to $\mathbf{excl}\,(\mathrm{at}\ L_1, \mathrm{at}\ L_2)$.)

Derivation of the assertion. Let $\bar{L}_1 = L_1 \cup \{\alpha_2\}$, $\bar{L}_2 = L_2 \cup \{\beta_2\}$ and $A \equiv \mathbf{excl}(\mathrm{at}\ \bar{L}_1, \mathrm{at}\ \bar{L}_2, ex = \mathbf{true})$. We show that A is an appropriate invariant for use of the rule (inv). Since A also implies $\mathbf{excl}\,(\mathrm{at}\ L_1, \mathrm{at}\ L_2)$ according to the above remarks we then get the assertion. Formally the derivation runs as follows:

(1)	$\mathrm{start}_\Pi \to A$	definition of start_Π
(2)	$A\ \mathbf{invof}\ \mathcal{M}_\Pi \setminus \{\alpha_1, \alpha_2, \beta_1, \beta_2\}$	since without occurrence of ex
(3)	$\alpha_1 \to ex = \mathbf{true} \wedge \circ\, ex = \mathbf{false}$	(Π)
(4)	$\alpha_1 \wedge A \to \neg\ \mathrm{at}\ \bar{L}_1 \wedge \neg\ \mathrm{at}\ \bar{L}_2$	(3)
(5)	$\alpha_1 \wedge A \to \circ A$	(Π), (3), (4)
(6)	$\alpha_2 \to \mathrm{at}\ \bar{L}_1 \wedge \circ\neg\ \mathrm{at}\ \bar{L}_1$	(Π)
(7)	$\alpha_2 \wedge A \to \neg\ \mathrm{at}\ \bar{L}_2 \wedge ex \neq \mathbf{true}$	(6)
(8)	$\alpha_2 \wedge A \to \circ A$	(Π), (6), (7)
(9)	$A\ \mathbf{invof}\ \{\beta_1, \beta_2\}$	in the same way as (5), (8)
(10)	$A\ \mathbf{invof}\ \mathcal{M}_\Pi$	(2), (5), (8), (9)
(11)	$\mathrm{start}_\Pi \to \Box A$	(inv), (1), (10)
(12)	$A \to \mathbf{excl}\,(\mathrm{at}\ L_1, \mathrm{at}\ L_2)$	(taut)
(13)	$\mathrm{start}_\Pi \to \Box\ \mathbf{excl}\,(\mathrm{at}\ L_1, \mathrm{at}\ L_2)$	(11), (12) ☐

Observe that we now begin to feel free to shorten our comments on how some line in a derivation is precisely derived. We omit the listing of the purely logical rules and laws like (prop) or (T1), (T2), ... This is a first step to a more feasible use of the formal logical apparatus. It presupposes some experience and practice with the logic but a user who has achieved a certain safety in formal reasoning will be able to proceed in a somewhat less stringent manner, maybe even comparable to the way in which a mathematician who is sufficiently sure about classical logic does not really derive his or her theorems within a formal proof system. We do not want to go so far and we will continue to write our proofs as formal derivations, but we will condense simple logical arguments and also separate them from the main proof ideas involved.

17. Examples of Applications

In this section we illustrate the use of the invariant method for invariance properties by three somewhat more complicated examples. We begin with a "reader/writer" program in which a more refined mutual exclusion pattern is realized.

Program Example 17.1

$$\psi_1 \equiv \textbf{initial}\ ex = \textbf{true} \wedge s = \textbf{true} \wedge num = 0;$$
$$\textbf{cobegin}\ \Pi_1^r \parallel \ldots \parallel \Pi_n^r \parallel \Pi_1^w \parallel \ldots \parallel \Pi_m^w\ \textbf{coend}$$

Every Π_i^r, $1 \le i \le n$ is a "reader" of the form:

```
Π_i^r ≡ loop α_0^(i): await ex=true then ex:=false;
             α_1^(i): num:=num+1;
             α_2^(i): if num=1 then
                        α_3^(i): await s=true then s:=false fi;
             α_4^(i): ex:=true;
                        :}read section
             α_5^(i): await ex=true then ex:=false;
             α_6^(i): num:=num-1;
             α_7^(i): if num=0 then α_8^(i): s:=true fi;
             α_9^(i): ex:=true
        end
```

and every Π_j^w, $1 \le j \le m$, is a "writer" of the form

```
Π_j^w ≡ loop β_0^(j): await s=true then s:=false;
                        :}write section
             β_1^(j): s:=true
        end
```

(without any occurrences of *ex*, *s*, and *num* in the read and write sections).

The synchronization in these program components is one possible solution for achieving the following goal:

- at most one writer may be in its write section,
- writers and readers may not be in their write and read sections at the same time.
  (However, arbitrarily many readers may be in their read sections at the same time.)

The essential idea in this solution is to count the number of "active" readers on the variable *num*. The exact functioning will be made clear by the subsequent proof.

If we take the notation:

$$L_i^r = \text{set of labels in the read section of } \Pi_i^r, \qquad i = 1, \ldots, n,$$
$$L_j^w = \text{set of labels in the write section of } \Pi_j^w, \qquad j = 1, \ldots, m$$

then we can first formulate that ψ_1 realize the above goal:

Assertion

$\psi_1 \vdash \text{start}_{\psi_1} \rightarrow \Box\ \textbf{excl}\ ((\text{at } L_1^r \vee \ldots \vee \text{at } L_n^r), \text{at } L_1^w, \ldots, \text{at } L_m^w)$.

To prove this we again introduce some abbreviations. First it is clear that we can define *k*-ary propositional operators **exor0**, **exor1**, **exor2**, etc., with:

$$\textbf{exori}\,(A_1, \ldots, A_k)$$

meaning "exactly i formulas out of $A_1, \ldots, A_k$ are true". For example,

$$\mathbf{exor0}(A_1, \ldots, A_k) \equiv \neg A_1 \wedge \ldots \wedge \neg A_k,$$
$$\mathbf{exor1}(A_1, \ldots, A_k) \equiv \mathbf{exor}(A_1, \ldots, A_k),$$
etc.

Then we write:

$$A_1 + A_2 + \ldots + A_k = l \quad \text{for } (\mathbf{exor0}(A_1, \ldots, A_k) \rightarrow l=0) \wedge (\mathbf{exor1}(A_1, \ldots, A_k) \rightarrow l=1) \wedge \ldots (\mathbf{exork}(A_1, \ldots, A_k) \rightarrow l=k)$$

informally meaning "l = number of those $A_i \in \{A_1, \ldots, A_k\}$ which are true". Finally we use the following sets of labels for $i=1, \ldots, n$ and $j=1, \ldots, m$:

$$K_i = L_i^r \cup \{\alpha_4^{(i)}, \alpha_5^{(i)}, \alpha_6^{(i)}, \alpha_7^{(i)}, \alpha_8^{(i)}\},$$
$$\bar{K}_i = L_i^r \cup \{\alpha_2^{(i)}, \alpha_3^{(i)}, \alpha_4^{(i)}, \alpha_5^{(i)}, \alpha_6^{(i)}\},$$
$$\bar{L}_i^r = L_i^r \cup \{\alpha_5^{(i)}\},$$
$$\bar{L}_j^w = L_j^w \cup \{\beta_1^{(j)}\},$$
$$M_i = \{\alpha_1^{(i)}, \alpha_2^{(i)}, \alpha_3^{(i)}, \alpha_4^{(i)}\},$$
$$N_i = \{\alpha_6^{(i)}, \alpha_7^{(i)}, \alpha_8^{(i)}, \alpha_9^{(i)}\}.$$

Proof of the assertion. The binary semaphore *ex* ensures mutually exclusive access to the variable *num*. A completely analogous proof as in the previous section shows that:

(1) $\text{start}_{\psi_1} \rightarrow \Box\, \mathbf{excl}\,(\text{at } M_1, \ldots, \text{at } M_n, \text{at } N_1, \ldots, \text{at } N_n)$.

Next we show that "the value of *num* is always the number of readers which are between their α_1 and α_7". Formally:

(2) $\text{start}_{\psi_1} \rightarrow \Box\,(\text{at } \bar{K}_1 + \ldots + \text{at } \bar{K}_n = num)$.

Derivation of (2). Let $A \equiv (\text{at } \bar{K}_1 + \ldots + \text{at } \bar{K}_n = num)$.

(2.1) $\text{start}_{\psi_1} \rightarrow A$ — definition of start

(2.2) $A\ \mathbf{invof}\ \mathcal{M}_{\psi_1} \setminus \bigcup_{i=1}^{n} \{\alpha_1^{(i)}, \alpha_6^{(i)}\}$ — (Π)

(2.3) $A\ \mathbf{invof}\ \bigcup_{i=1}^{n} \alpha_1^{(i)}$ — (Π)

(2.4) $A\ \mathbf{invof}\ \bigcup_{i=1}^{n} \alpha_6^{(i)}$ — (Π)

(2.5) $\text{start}_{\psi_1} \rightarrow \Box A$ — (inv), (2.1)–(2.4)

With (1) and (2) we can show: "If a reader finds $num \neq 1$ at α_2, then at least one (other) reader is at α_5 or in the read section", i.e.,

(3) $\text{start}_{\psi_1} \rightarrow \Box\,(\text{at } \alpha_2^{(i)} \wedge num \neq 1 \rightarrow \text{at } \bar{L}_1^r \vee \ldots \vee \text{at } \bar{L}_n^r)$ (for every i).

Derivation of (3)

(3.1)	$\square(\text{at } \alpha_2^{(i)} \rightarrow num \geq 1)$	(2)
(3.2)	$\square(\text{at } \alpha_2^{(i)} \wedge num \neq 1 \rightarrow num \geq 2)$	(3.1)
(3.3)	$\square(\text{at } \alpha_2^{(i)} \wedge num \neq 1 \rightarrow \text{at } \bar{K}_1 \vee \ldots \vee \text{ at } \bar{K}_{i-1} \vee \text{at } \bar{K}_{i+1} \vee \ldots \vee \text{at } \bar{K}_n)$	(2), (3.2)
(3.4)	$\square(\text{at } \alpha_2^{(i)} \rightarrow \neg \text{ at } M_k)$ for every $k=1, \ldots, n, k \neq i$	(1)
(3.5)	$\square(\text{at } \alpha_2^{(i)} \rightarrow \neg \text{ at } N_k)$ for every $k=1, \ldots, n$	(1)
(3.6)	$\square(\text{at } \alpha_2^{(i)} \wedge num \neq 1 \rightarrow \text{at } \bar{L}_1^r \vee \ldots \vee \text{at } \bar{L}_{i-1}^r \vee \text{at } \bar{L}_{i+1}^r \vee \ldots \vee \text{at } \bar{L}_n^r)$	(3.3)–(3.5)
(3.7)	$\text{start}_{\psi_1} \rightarrow \square(\text{at } \alpha_2^{(i)} \wedge num \neq 1 \rightarrow \text{at } \bar{L}_1^r \vee \ldots \vee \text{at } \bar{L}_n^r)$	(3.6)

Next we consider the statement α_8. We claim that "if a reader is at α_8 then $num=0$", i.e.,

(4) $\text{start}_{\psi_1} \rightarrow \square(\text{at } \alpha_8^{(i)} \rightarrow num=0)$ (for every i).

Derivation of (4). Let $M = \bigcup_{k=1}^{n} \{\alpha_1^{(k)}, \alpha_6^{(k)}, \alpha_7^{(k)}\}$.

(4.1)	$\text{start}_{\psi_1} \rightarrow (\text{at } \alpha_8^{(i)} \rightarrow num=0)$	definition of start_{ψ_1}
(4.2)	$(\text{at } \alpha_8^{(i)} \rightarrow num=0)$ **invof** $\bar{\mathcal{M}}_\Pi \backslash M$	(Π)
(4.3)	$\alpha \rightarrow \bigcirc \neg \text{ at } \alpha_8^{(i)}$ for every $\alpha \in M \backslash \{\alpha_7^{(i)}\}$	(Π), (1)
(4.4)	$(\text{at } \alpha_8^{(i)} \rightarrow num=0)$ **invof** $M \backslash \{\alpha_7^{(i)}\}$	(4.3)
(4.5)	$(\text{at } \alpha_8^{(i)} \rightarrow num=0)$ **invof** $\alpha_7^{(i)}$	(Π)
(4.6)	$\text{start}_{\psi_1} \rightarrow \square(\text{at } \alpha_8^{(i)} \rightarrow num=0)$	(inv), (4.1), (4.2), (4.4), (4.5)

With (4) we can show that "if a reader is at α_8 then no other reader is in the section given by the label set K", i.e.,

(5) $\text{start}_{\psi_1} \rightarrow \square(\text{at } \alpha_8^{(i)} \rightarrow \neg \text{at } K_1 \wedge \ldots \wedge \neg \text{ at } K_{i-1} \wedge \neg \text{ at} K_{i+1} \wedge \ldots \wedge \neg \text{ at } K_n)$ (for every i).

Derivation of (5)

(5.1)	$\square(\text{at } \alpha_8^{(i)} \rightarrow num=0)$	(4)
(5.2)	$\square(\text{at } \alpha_8^{(i)} \rightarrow \neg \text{ at } \bar{K}_1 \wedge \ldots \wedge \neg \text{ at } \bar{K}_n)$	(2), (5.1)
(5.3)	$\square(\text{at } \alpha_8^{(i)} \rightarrow \neg \text{ at } M_k \wedge \neg \text{ at } N_k)$ for $k=1, \ldots, n, k \neq i$	(1)
(5.4)	$\text{start}_{\psi_1} \rightarrow \square(\text{at } \alpha_8^{(i)} \rightarrow \neg \text{ at } K_1 \wedge \ldots \wedge \neg \text{ at } K_{i-1} \wedge \neg \text{ at } K_{i+1} \wedge \ldots \wedge \neg \text{ at } K_n)$	(5.2), (5.3)

Now let

$$A \equiv \textbf{excl}((\text{at } K_1 \vee \ldots \vee \text{at } K_n), \text{at } \bar{L}_1^w, \ldots, \text{at } \bar{L}_m^w, s=\textbf{true}).$$

It is not difficult to see:

(6) $\text{start}_{\psi_1} \rightarrow A$,

(7) A **invof** $\bar{\mathcal{M}}_{\psi_1} \backslash \bigcup_{i=1}^{n} \{\alpha_2^{(i)}, \alpha_8^{(i)}\}$.

We show that A is also an invariant of all $\alpha_2^{(i)}$ and $\alpha_8^{(i)}$:

(8) A **invof** $\alpha_2^{(i)}$ (for every i),

(9) A **invof** $\alpha_8^{(i)}$ (for every i).

Derivation of (8)

(8.1)	$\alpha_2^{(i)} \wedge num=1 \wedge A \rightarrow \circ A$	(Π)
(8.2)	$\alpha_2^{(i)} \wedge num \neq 1 \rightarrow \text{at } K_1 \vee \ldots \vee \text{at } K_n$	(3)
(8.3)	$\alpha_2^{(i)} \wedge num \neq 1 \wedge A \rightarrow \circ A$	(8.2)
(8.4)	A **invof** $\alpha_2^{(i)}$	(8.1), (8.3)

Derivation of (9)

(9.1)	$\alpha_8^{(i)} \rightarrow \neg \text{ at } K_1 \wedge \ldots \wedge$ $\neg \text{ at } K_{i-1} \wedge \text{at } K_i \wedge \neg \text{ at } K_{i+1} \wedge \ldots \wedge \neg \text{ at } K_n$	(5)
(9.2)	$\alpha_8^{(i)} \wedge A \rightarrow \neg \text{ at } \bar{L}_1^w \wedge \ldots \wedge \neg \text{ at } \bar{L}_m^w$	(9.1)
(9.3)	$\alpha_8^{(i)} \wedge A \rightarrow \circ(\neg (\text{at } K_1 \vee \ldots \vee \text{at } K_n) \wedge \neg \text{ at } \bar{L}_1^w \wedge \ldots \wedge \neg \text{ at } \bar{L}_m^w)$	(Π), (9.1), (9.2)
(9.4)	A **invof** $\alpha_8^{(i)}$	(9.3)

From (6)–(9) we finally get, by (inv):

(10) $\text{start}_{\psi_1} \rightarrow \Box A,$

and since we clearly have:

(11) $A \rightarrow$ **excl** $((\text{at } L_1^r \vee \ldots \vee \text{at } L_n^r), \text{at } L_1^w, \ldots, \text{at } L_m^w)$

we also get the desired result:

(12) $\text{start}_{\psi_1} \rightarrow \Box$ **excl** $((\text{at } L_1^r \vee \ldots \vee \text{at } L_n^r), \text{at } L_1^w, \ldots, \text{at } L_m^w)$. □

Our next example deals with the partial correctness of a non-cyclic program.

Program Example 17.2

```
ψ₂ ≡ initial i=1 ∧ rset=empty ∧ k₁=1 ∧ k₂=1 ∧ a=0 ∧ b=0 ∧ n≥0;
    cobegin α₀ : while k₁≤n do
                    α₁ : a:=i;   i:=i+1;
                    α₂ : if p(a) then append(rset, a) fi;
                    α₃ : k₁:=k₁+1 od;
                α₄ : stop
                ||
                β₀ : while k₂≤n do
                    β₁ : b:=i;   i:=i+1;
                    β₂ : if p(b) then append(rset, b) fi;
                    β₃ : k₂:=k₂+1 od;
                β₄ : stop
    coend
```

Each parallel component of ψ_2 runs n times through its loop. In every run it gets a new value of i on its private variable a or b, respectively, and tests whether this i fulfills a predicate p. If it does, the value is stored in the "result set" *rset*. Upon termination of ψ_2, *rset* should contain exactly those $i \in \{1, \ldots, 2n\}$ with $p(i)$ true.

In ψ_2, rather compound statements are turned into elementary ones. As indicated earlier we presuppose an appropriate synchronization. The "effects" of α_1, α_2, β_1 and β_2 are informally clear. We only indicate a formal specification for the action

α_2. It can be axiomatized by the following three formulas:

- $\alpha_2 \wedge p(a) \wedge rset = rset_0 \rightarrow \circ(rset = rset_0 \cup \{a\})$,
- $\alpha_2 \wedge p(a) \wedge P \rightarrow \circ P$ for every P-formula P not containing *rset*,
- $\alpha_2 \wedge \neg p(a) \wedge P \rightarrow \circ P$ for every P-formula P.

Now let

$$A \equiv \forall x(x \in rset \rightarrow p(x) \wedge x \leq 2n),$$
$$B \equiv \forall x(p(x) \wedge x \leq 2n \rightarrow x \in rset).$$

A and B describe the desired result, so we can formulate the assertion for ψ_2.

Assertion

$$\psi_2 \vdash \text{start}_{\psi_2} \rightarrow \Box(\text{at } \alpha_4 \wedge \text{at } \beta_4 \rightarrow A \wedge B).$$

We do not carry out the proof of this assertion in all details. We only give the main steps of one "half" of it leaving the rest as an exercise to the reader.

Proof of $\text{start}_{\psi_2} \rightarrow \Box(\text{at } \alpha_4 \wedge \text{at } \beta_4 \rightarrow A)$
Taking over notation from Example 17.1 we first find by trivial applications of (inv):

(1) $\text{start}_{\psi_2} \rightarrow \Box(\text{at } \{\alpha_0, \alpha_1, \alpha_4\} + \text{at } \{\beta_0, \beta_1, \beta_4\} = k_1 + k_2 + 1 - i)$
(2) $\text{start}_{\psi_2} \rightarrow \Box(\text{at } \{\alpha_0, \alpha_4\} \rightarrow k_1 \leq n+1) \wedge (\neg \text{at } \{\alpha_0, \alpha_4\} \rightarrow k_1 \leq n)$
(3) $\text{start}_{\psi_2} \rightarrow \Box(\text{at } \{\beta_0, \beta_4\} \rightarrow k_2 \leq n+1) \wedge (\neg \text{at } \{\beta_0, \beta_4\} \rightarrow k_2 \leq n)$

From this we get:

(4) $\text{start}_{\psi_2} \rightarrow \Box(\text{at } \alpha_1 \rightarrow i \leq 2n)$
(5) $\text{start}_{\psi_2} \rightarrow \Box(\text{at } \beta_1 \rightarrow i \leq 2n)$

We give only the

Derivation of (4)

(4.1)	$\Box(\text{at } \alpha_1 \wedge \text{at } \{\beta_0, \beta_4\} \rightarrow i = k_1 + k_2 - 1)$	(1)
(4.2)	$\Box(\text{at } \alpha_1 \wedge \text{at } \{\beta_0, \beta_4\} \rightarrow i \leq 2n)$	(2), (3), (4.1)
(4.3)	$\Box(\text{at } \alpha_1 \wedge \neg \text{ at } \{\beta_0, \beta_4\} \rightarrow i \leq k_1 + k_2)$	(1)
(4.4)	$\Box(\text{at } \alpha_1 \wedge \neg \text{ at } \{\beta_0, \beta_4\} \rightarrow i \leq 2n)$	(2), (3), (4.3)
(4.5)	$\text{start}_{\psi_2} \rightarrow \Box(\text{at } \alpha_1 \rightarrow i \leq 2n)$	(4.2), (4.4)

From (4) and (5) we deduce quite trivially:

(6) $\text{start}_{\psi_2} \rightarrow \Box(a \leq 2n \wedge b \leq 2n)$

and from this we get:

(7) $\text{start}_{\psi_2} \rightarrow \Box A$

which shows that A is even a global invariant and therefore holds also upon termination: $\text{start}_{\psi_2} \rightarrow \Box(\text{at } \alpha_4 \wedge \text{at } \beta_4 \rightarrow A)$.

Derivation of (7)

(7.1)	$\mathrm{start}_{\psi_2} \rightarrow A$	definition of start_{ψ_2}
(7.2)	A **invof** α_2	(Π), (6)
(7.3)	A **invof** β_2	(Π), (6)
(7.4)	A **invof** $\{\alpha_0, \alpha_1, \alpha_3, \beta_0, \beta_1, \beta_3\}$	trivial
(7.5)	$\mathrm{start}_{\psi_2} \rightarrow \Box A$	(inv), (7.1)–(7.4) □

Our third example deals with deadlock freedom. We consider the well-known producer-consumer-with-bounded-buffer program.

Program Example 17.3

```
ψ3 ≡ initial ex=true ∧ bf=0 ∧ be=n ∧ n>0;
     cobegin loop α0 : ~ } produce section
                       :
                       :
                  α1 : await be>0 then be:=be−1;
                  α2 : await ex=true then ex:=false;
                       :}store section
                  α3 : ex:=true;
                  α4 : bf:=bf+1
             end
             ||
             loop β0 : await bf>0 then bf:=bf−1;
                  β1 : await ex=true then ex:=false;
                       :}get section
                  β2 : ex:=true;
                  β3 : be:=be+1;
                       :}consume section
              end
     coend
```

The first component of ψ_3 – the producer – produces an object in every loop run and stores it in a shared buffer which can contain up to n such objects. The other component – the consumer – periodically gets an object out of the buffer and consumes it. Storing in the buffer presupposes that it is not full ($be>0$) and getting something from the buffer presupposes that it is not empty ($bf>0$). Variables like be and bf (in this case counting the "empty" and "full" lots of the buffer) with operations as given are called *general semaphores*. The mutual exclusion of access to the buffer is again synchronized by a binary semaphore ex. The order of the synchronization statements is very important. If, for example, α_1 and α_2 were exchanged then a deadlock could occur. The given synchronization, however, is free of deadlocks:

Assertion

a) $\psi_3 \vdash \mathrm{start}_{\psi_3} \rightarrow \Box\,(\mathrm{at}\ \alpha_1 \wedge \mathrm{at}\ \beta_0 \rightarrow be>0 \vee bf>0)$.

b) $\psi_3 \vdash \mathrm{start}_{\psi_3} \rightarrow \Box\,(\mathrm{at}\ \alpha_1 \wedge \mathrm{at}\ \beta_1 \rightarrow be>0 \vee ex=\mathbf{true})$.

c) $\psi_3 \vdash \text{start}_{\psi_3} \to \Box(\text{at } \alpha_2 \wedge \text{at } \beta_0 \to ex = \mathbf{true} \vee bf > 0)$.
d) $\psi_3 \vdash \text{start}_{\psi_3} \to \Box(\text{at } \alpha_2 \wedge \text{at } \beta_1 \to ex = \mathbf{true})$.

We define the following sets of labels:

L_1 = set of labels in store section $\cup \{\alpha_3\}$,
L_2 = set of labels in get section $\cup \{\beta_2\}$,
M_1 = set of labels in produce section $\cup \{\alpha_1\}$,
M_2 = set of labels in consume section $\cup \{\beta_0\}$,

and, again, take over earlier notations.

Proof of the assertion. By a proof very similar to that in Example 17.1 one can show:

(1) $\text{start}_{\psi_3} \to \Box\, \mathbf{exor}(\text{at } L_1, \text{at } L_2, ex = \mathbf{true})$.

From (1), parts b), c) and d) of the assertion follow immediately with (prop). For the proof of a) we first note:

(2) $\text{start}_{\psi_3} \to \Box(be \geq 0 \wedge bf \geq 0)$

which is trivially found using (inv), and with (2) we find

(3) $\text{start}_{\psi_3} \to \Box(\text{at } M_1 \wedge \text{at } M_2 \to be > 0 \vee bf > 0)$.

From (3), part a) follows immediately.

Derivation of (3). Let $A \equiv \text{at } M_1 \wedge \text{at } M_2 \to be > 0 \vee bf > 0$.

(3.1)	$\text{start}_{\psi_3} \to A$	definition of start_{ψ_3}
(3.2)	$\alpha_1 \to \bigcirc \neg \text{at } M_1$	(Π)
(3.3)	A **invof** α_1	(3.2)
(3.4)	A **invof** β_0	as (3.3)
(3.5)	$\alpha_4 \to bf \geq 0$	(2)
(3.6)	$\alpha_4 \to \bigcirc(bf > 0)$	(Π), (3.5)
(3.7)	A **invof** α_4	(3.6)
(3.8)	A **invof** β_3	as (3.7)
(3.9)	A **invof** $\mathscr{M}_{\psi_3} \setminus \{\alpha_1, \alpha_4, \beta_0, \beta_3\}$	trivial
(3.10)	$\text{start}_{\psi_3} \to \Box A$	(inv), (3.1), (3.3), (3.4), (3.7), (3.8), (3.9) □

18. Invariant Methods for Precedence Properties

Precedence properties of a program Π are expressed by formulas using operators **atnext**, **unless**, etc. The basic form of such formulas is:

$A \to B$ **atnext** C

or, in the same way, with one of the other operators. However, we have already seen in Section 15 that we should also consider slight modifications to this form, e.g.,

$$A \rightarrow B \ \mathbf{atnext} \ C \vee B \ \mathbf{atnext}^2 \ C.$$

Again we remember general logical proof rules for such formulas from Section 7, for example,

(indatnext) $A \rightarrow \circ(C \rightarrow B) \wedge \circ(\neg C \rightarrow A) \vdash A \rightarrow B \ \mathbf{atnext} \ C.$

The transfer of such a rule to the situation where programs are involved runs as it does with the rule (ind″) in Section 16, using (trans). In the case of (indatnext) we are led to the following "invariant" rule for (simple) atnext formulas:

(atnext) $\alpha \wedge A \rightarrow \circ(C \rightarrow B) \wedge \circ(\neg C \rightarrow A)$ for every $\alpha \in \bar{\mathscr{M}}_\Pi$,
$nil_\Pi \wedge A \rightarrow (C \rightarrow B)$
$\vdash A \rightarrow B \ \mathbf{atnext} \ C.$

Formula A in this rule is a kind of invariant which is preserved by every action of Π, at least as long as C does not become true.

Theorem 18.1. *The rule* (*atnext*) *is* Π*-derived.*

Proof. We give the simple derivation:

(1)	$\alpha \wedge A \rightarrow \circ(C \rightarrow B) \wedge \circ(\neg C \rightarrow A)$ for every $\alpha \in \bar{\mathscr{M}}_\Pi$	assumption
(2)	$nil_\Pi \wedge A \rightarrow (C \rightarrow B)$	assumption
(3)	$nil_\Pi \wedge A \rightarrow (C \rightarrow B) \wedge (\neg C \rightarrow A)$	(prop), (2)
(4)	$A \rightarrow \circ(C \rightarrow B) \wedge \circ(\neg C \rightarrow A)$	(trans), (1), (3)
(5)	$A \rightarrow B \ \mathbf{atnext} \ C$	(indatnext), (4) ☐

Obviously we may deduce in exactly the same way invariant rules for the operators **unless**, **while** and **before** from the induction principles (indunless), (indwhile) and (indbefore), respectively:

(unless) $\alpha \wedge A \rightarrow \circ C \vee \circ(A \wedge B)$ for every $\alpha \in \bar{\mathscr{M}}_\Pi$,
$nil_\Pi \wedge A \rightarrow B \vee C$
$\vdash A \rightarrow B \ \mathbf{unless} \ C$

(while) $\alpha \wedge A \rightarrow \circ(C \rightarrow A \wedge B)$ for every $\alpha \in \bar{\mathscr{M}}_\Pi$,
$nil_\Pi \wedge A \rightarrow (C \rightarrow B)$
$\vdash A \rightarrow B \ \mathbf{while} \ C$

(before) $\alpha \wedge A \rightarrow \circ\neg C \wedge \circ(A \vee B)$	for every $\alpha \in \bar{\mathcal{M}}_\Pi$,
$\mathrm{nil}_\Pi \wedge A \rightarrow \neg C$	
$\vdash A \rightarrow B$ **before** C	

We give a simple example of how to use such rules, in this case the rule (atnext).

Example. Let

$$
\begin{aligned}
\Pi \equiv\ & \texttt{initial}\ k=3;\\
& \texttt{loop}\ \ \alpha_0: \texttt{if}\ \mathrm{prime}(k)\ \texttt{then}\\
& \qquad\qquad\qquad \alpha_1: \mathrm{output}(k)\ \texttt{fi};\\
& \qquad\quad \alpha_2:\ k := k+2\\
& \texttt{end}
\end{aligned}
$$

be a sequential program of the kind indicated in Section 15. Π is to output the sequence of odd prime numbers. We put:

$$\mathrm{nextprime}(n) := \text{smallest prime number} > n$$

and formulate the desired effect of Π by:

a) $\Pi \vdash \mathrm{start}_\Pi \rightarrow k=3$ **atnext** at α_1,
b) $\Pi \vdash \mathrm{at}\ \alpha_1 \wedge k=k_0 \rightarrow k=\mathrm{nextprime}(k_0)$ **atnext** at α_1.

For the proof of a) we see that $\mathrm{start}_\Pi \rightarrow \circ(\mathrm{at}\ \alpha_1 \wedge k=3)$ holds, and hence we have $\mathrm{start}_\Pi \rightarrow \circ(\mathrm{at}\ \alpha_1 \rightarrow k=3) \wedge \circ(\neg\mathrm{at}\ \alpha_1 \rightarrow \mathrm{start}_\Pi)$. Direct application of (indatnext) yields the assertion a). For the proof of b) we first note some simple arithmetical properties of the function nextprime:

(1) $k > k_0 \wedge k \le \mathrm{nextprime}(k_0) \wedge \mathrm{prime}(k) \rightarrow k = \mathrm{nextprime}(k_0)$
(2) $k \le \mathrm{nextprime}(k_0) \wedge \neg\mathrm{prime}(k) \rightarrow k < \mathrm{nextprime}(k_0)$
(3) $k_0 < \mathrm{nextprime}(k_0)$
(4) $k < \mathrm{nextprime}(k_0) \wedge \mathrm{odd}(k) \wedge k \ge 3 \rightarrow k+2 \le \mathrm{nextprime}(k_0)$.

Furthermore, we have the trivial global invariants:

(5) $\Box(\mathrm{odd}(k) \wedge k \ge 3)$
(6) $\Box\neg\mathrm{nil}_\Pi$

which are simply shown with (inv′). The informal reason for (6) is that Π contains no end labels and no await statements.

We now want to apply rule (atnext), but before doing so we still have to specify formally the effect of the statement output(k). Certainly this statement is to change no variables, and therefore we may asssume as an additional specification axiom:

(7) $\alpha_1 \wedge P \rightarrow \circ P$ for every P-formula P.

Now let

$$
\begin{aligned}
A \equiv\ & (\mathrm{at}\ \alpha_0 \rightarrow k > k_0 \wedge k \le \mathrm{nextprime}(k_0)) \wedge\\
& (\mathrm{at}\ \alpha_1 \rightarrow k = k_0) \wedge\\
& (\mathrm{at}\ \alpha_2 \rightarrow k \ge k_0 \wedge k < \mathrm{nextprime}(k_0)).
\end{aligned}
$$

Then we have:

(8)	$\alpha_0 \wedge A \wedge \text{prime}(k) \rightarrow \circ(\text{at } \alpha_1 \wedge k = \text{nextprime}(k_0))$	(Π), (1)
(9)	$\alpha_0 \wedge A \wedge \neg\text{prime}(k) \rightarrow \circ(\text{at } \alpha_2 \wedge k \geq k_0 \wedge k < \text{nextprime}(k_0))$	(Π), (2)
(10)	$\alpha_0 \wedge A \rightarrow \circ(\text{at } \alpha_1 \rightarrow k = \text{nextprime}(k_0)) \wedge \circ(\neg \text{at } \alpha_1 \rightarrow A)$	(8), (9)
(11)	$\alpha_1 \wedge A \rightarrow \circ(\text{at } \alpha_2 \wedge k \geq k_0 \wedge k < \text{nextprime}(k_0))$	(Π), (3), (7)
(12)	$\alpha_1 \wedge A \rightarrow \circ(\text{at } \alpha_1 \rightarrow k = \text{nextprime}(k_0)) \wedge \circ(\neg \text{at } \alpha_1 \rightarrow A)$	(11)
(13)	$\alpha_2 \wedge A \rightarrow \circ(\text{at } \alpha_0 \wedge k > k_0 \wedge k \leq \text{nextprime}(k_0))$	(Π), (4), (5)
(14)	$\alpha_2 \wedge A \rightarrow \circ(\text{at } \alpha_1 \rightarrow k = \text{nextprime}(k_0)) \wedge \circ(\neg \text{at } \alpha_1 \rightarrow A)$	(13)
(15)	$\text{nil}_\Pi \wedge A \rightarrow (\text{at } \alpha_1 \rightarrow k = \text{nextprime}(k_0))$	(6)
(16)	$A \rightarrow k = \text{nextprime}(k_0)$ **atnext** at α_1	(atnext), (10), (12), (14), (15)

Since at $\alpha_1 \wedge k = k_0 \rightarrow A$ we get assertion b) directly from (16). □

Turning now to modified formulas as indicated above, we recall induction principles for the iterated atnext operator, e.g.,

(indatnext²) $A \rightarrow \circ(C \rightarrow B_1) \wedge \circ(\neg C \rightarrow A)$,
$B_1 \rightarrow \circ(C \rightarrow B) \wedge \circ(\neg C \rightarrow B_1)$
$\vdash A \rightarrow B$ **atnext²** C.

Again, we can transfer this to:

(atnext²) $\alpha \wedge A \rightarrow \circ(C \rightarrow B_1) \wedge \circ(\neg C \rightarrow A)$ for every $\alpha \in \bar{\mathcal{M}}_\Pi$,
$\text{nil}_\Pi \wedge A \rightarrow (C \rightarrow B_1)$,
$\alpha \wedge B_1 \rightarrow \circ(C \rightarrow B) \wedge \circ(\neg C \rightarrow B_1)$ for every $\alpha \in \bar{\mathcal{M}}_\Pi$,
$\text{nil}_\Pi \wedge B_1 \rightarrow (C \rightarrow B)$
$\vdash A \rightarrow B$ **atnext²** C.

The generalization to (atnext^n) for $n > 2$ should be clear.

A closer look at (indatnext²) even shows that the third and fourth premises of this rule are just premises of (indatnext), so we may also infer $B_1 \rightarrow B$ **atnext** C (compare with the proof of (indatnext²) in Section 7). This shows that modified implications like the one at the beginning of this section (containing **atnext** and **atnext²**) can be derived by simply extending the conclusion in (atnext²) or, generally, in (atnext^n):

(disatnext^n) $\alpha \wedge A \rightarrow \circ(C \rightarrow B_1) \wedge \circ(\neg C \rightarrow A)$ for every $\alpha \in \bar{\mathcal{M}}_\Pi$,
$\alpha \wedge B_1 \rightarrow \circ(C \rightarrow B_2) \wedge \circ(\neg C \rightarrow B_1)$ for every $\alpha \in \bar{\mathcal{M}}_\Pi$,
⋮
$\alpha \wedge B_{n-1} \rightarrow \circ(C \rightarrow B) \wedge \circ(\neg C \rightarrow B_{n-1})$ for every $\alpha \in \bar{\mathcal{M}}_\Pi$,
$\text{nil}_\Pi \wedge A \rightarrow (C \rightarrow B_1)$,
$\text{nil}_\Pi \wedge B_1 \rightarrow (C \rightarrow B_2)$,
⋮
$\text{nil}_\Pi \wedge B_{n-1} \rightarrow (C \rightarrow B)$
$\vdash A \vee B_1 \vee \ldots \vee B_{n-1} \rightarrow B$ **atnext** $C \vee \ldots \vee B$ **atnext**^{n} C

Theorem 18.2. *The rule (disatnext^n) is Π-derived.*

Proof. We show only the case $n=2$. The generalization is trivial.

(1)	$\alpha \wedge A \rightarrow \circ(C \rightarrow B_1) \wedge \circ(\neg C \rightarrow A)$	for every $\alpha \in \bar{\mathcal{M}}_\Pi$	assumption
(2)	$\alpha \wedge B_1 \rightarrow \circ(C \rightarrow B) \wedge \circ(\neg C \rightarrow B_1)$	for every $\alpha \in \bar{\mathcal{M}}_\Pi$	assumption
(3)	$nil_\Pi \wedge A \rightarrow (C \rightarrow B_1)$		assumption
(4)	$nil_\Pi \wedge B_1 \rightarrow (C \rightarrow B)$		assumption
(5)	$B_1 \rightarrow \circ(C \rightarrow B) \wedge \circ(\neg C \rightarrow B_1)$		(trans), (2), (4)
(6)	$B_1 \rightarrow B$ **atnext** C		(indatnext), (5)
(7)	$A \rightarrow B$ **atnext**^{2} C		(atnext^2), (1)–(4)
(8)	$A \vee B_1 \rightarrow B$ **atnext** $C \vee B$ **atnext**^{2} C		(prop), (6), (7) □

In the same way we could also find rules for iterated or nested applications of the other operators, e.g., for formulas of the kind:

$A \rightarrow B_1$ **unless** $[B_2$ **unless** $(B_3$ **unless** $C)]$

which occurred in Section 15. We leave the development of such rules as an exercise. We are now also able to explain the particular role of the basic program axioms (B1) and (B2) given in Section 14. All justifications of rules of this section make use only of axiom (B2) (via (trans)). Together with the analogous observation in Section 16 for the rule (inv) we may state that:

- the invariant methods for proving invariance and precedence (i.e., general safety) properties of programs are based on purely temporal logical laws, the basic rule (B1) and the basic axiom (B2).

This, of course, does not mean that the other program axioms are not really needed. They are applied – as already seen in the previous section – in verifying the various premises of the proof rules, i.e., they are applied if one concretely *uses* these rules.

19. Examples of Applications

Again we want to illustrate the use of the developed proof rules by three further examples. We begin with a program describing the principle of the so-called *alternating bit protocol.* This protocol is to solve the following problem. A "sender" transmits (in single steps) a sequence $U=u_0, u_1, u_2, \dots$ of "messages" to a "receiver". The receiver sends back acknowledgements but the transmissions in both directions may be corrupted by the "unreliable" transmission medium. Nevertheless, the correct messages should be "realized" by the receiver in the same order as they are sent (which includes, of course, without introducing repetitions). The main idea of the solution is to send messages together with a control bit, the value of which also serves as the acknowledgement and which has to alternate in an appropriate way.

Program Example 19.1

```
ψ4 ≡ initial nextinput=u0 ∧ nr=1 ∧ ls=mn=a=0;
     cobegin loop α0: if ls=a then ls:=ls⊕1;
                                   d:=nextinput fi;
                  α1: send(ls, d) to (mn, inf)
             end
             ||
             loop β0: if mn=nr then nextoutput:=inf;
                                    nr:=nr⊕1 fi;
                  β1: send(nr⊕1) to (a)
             end
     coend
```

The first component of ψ_4 describes the sender, the second one the receiver. The pair (mn, inf) of variables may be viewed as a "mail box" of the receiver and the variable a as a mail box of the sender. A send operation of the sender tries to write the control bit ls to mn and the actual message d to inf. This is repeated in the sender's loop with the same data until an appropriate acknowledgement has been realized. (Notice that this also means that a package (ls, d) correctly arrived on (mn, inf) may be corrupted again before the receiver has realized it.) The send operation β_1 is quite analogous. "Realizing" a correct message is implemented by the statement nextoutput$:=inf$. $\oplus$ denotes addition modulo 2.

A formal specification of the send operations α_1 and β_1 is given by the following additional axioms:

- $\alpha_1 \rightarrow \circ[(mn, inf)=(ls, d) \vee (mn, inf)=(\text{error}, \text{error})]$,
- $\alpha_1 \wedge P \rightarrow \circ P$ for every P-formula not containing mn and inf,
- $\beta_1 \rightarrow \circ(a=nr\oplus 1 \vee a=\text{error})$,
- $\beta_1 \wedge P \rightarrow \circ P$ for every P-formula not containing a.

The first and the third of these formulas describe the two possibilities of a correct or a corrupted transmission, the latter resulting in some (recognizable) error element in the mail box. The second and fourth formula express that no variables other than the relevant ones are changed.

Of course, we have also to specify formally the complex statements α_0 and β_0, but this should be obvious since α_0 and β_0 contain no unknown constructions. For example, we may assume the following axioms for α_0:

- $\alpha_0 \wedge ls=a \wedge \text{nextinput}=u_i \wedge ls=ls_0 \rightarrow \circ(\text{nextinput}=u_{i+1} \wedge ls=ls_0\oplus 1 \wedge d=u_i)$,
- $\alpha_0 \wedge P \rightarrow \circ P$ for every P-formula P not containing nextinput, ls, and d,
- $\alpha_0 \wedge ls \neq a \wedge P \rightarrow \circ P$ for every P-formula P,

and in the same way for β_0.

Observing now that the formula:

$$\beta_0 \wedge mn=nr$$

informally means that some message is realized by the receiver, the above-mentioned correct order of message realization is formalized by the following:

Assertion

a) $\psi_4 \vdash \text{start}_{\psi_4} \to \mathit{inf}=u_0$ **atnext** $(\beta_0 \wedge mn=nr)$.

b) $\psi_4 \vdash \beta_0 \wedge mn=nr \wedge \mathit{inf}=u_i \to \mathit{inf}=u_{i+1}$ **atnext** $(\beta_0 \wedge mn=nr)$.

It might be in order at this point to recall once more that both parts of this assertion do not contain any liveness property, i.e., they do not state that the messages really arrive. We will return to this point in Section 23. Here it has only been expressed that if messages arrive at all then their arrival and realization run correctly.

Proof of the assertion. The proofs of a) and b) use the same main ideas. We first note two trivial global invariants:

(1) $\Box((nr=0 \vee nr=1) \wedge (ls=0 \vee ls=1))$,

(2) $\Box(d=u_i \to \text{nextinput}=u_{i+1})$,

which are easily verified with (inv′). (We assume that initially d is undefined.) A more sophisticated invariant is:

(3) $\Box((mn=nr \to nr=ls \wedge \mathit{inf}=d) \wedge (ls=a \to nr \neq ls))$.

This holds by (inv′), since:

$$\text{start}_{\psi_4} \to mn \neq nr \wedge ls=a \wedge nr \neq ls$$

and $C \equiv (mn=nr \to nr=ls \wedge \mathit{inf}=d) \wedge (ls=a \to nr \neq ls)$ is left invariant by every $\gamma \in \bar{\mathcal{M}}_{\psi_4}$, e.g., for $\gamma \equiv \alpha_1$:

$$\alpha_1 \wedge C \to \circ[((mn, \mathit{inf})=(ls, d) \vee (mn, \mathit{inf})=(\text{error}, \text{error})) \wedge (ls=a \to nr \neq ls)],$$
$$(mn, \mathit{inf})=(ls, d) \to (mn=nr \to nr=ls \wedge \mathit{inf}=d),$$
$$mn=\text{error} \to mn \neq nr.$$

The last of these three formulas follows with (1). Together they yield:

$$\alpha_1 \wedge C \to \circ C.$$

Now let

$$B_i \equiv (nr=ls \to d=u_i) \wedge (nr \neq ls \to \text{nextinput}=u_i)$$

for every $i \in \mathbb{N}_0$. Again it is easy to check that:

(4) $\gamma \wedge B_i \to \circ B_i$ for $\gamma \in \{\alpha_0, \alpha_1, \beta_1\}$ and every $i \in \mathbb{N}_0$.

The actions α_1 and β_1 are trivial since B_i contains no mn, inf and a. For α_0 there is nothing to do in the case $ls \neq a$. In the case $ls=a$ we have:

$$\alpha_0 \wedge ls=a \wedge B_i \to \alpha_0 \wedge ls=a \wedge nr \neq ls \wedge \text{nextinput}=u_i$$

by (3) and:

$$\alpha_0 \wedge ls=a \wedge nr \neq ls \wedge \text{nextinput}=u_i \to \circ(nr=ls \wedge d=u_i)$$

with (1). We note further:

(5) $mn=nr \wedge B_i \rightarrow inf=u_i$

which follows immediately by (3) and are now ready to prove a) and b) with the rule (atnext). For a), we let:

$$A_a \equiv B_0 \wedge \neg(\beta_0 \wedge mn=nr).$$

Then we have:

(6) $\gamma \wedge A_a \rightarrow \circ(\beta_0 \wedge mn=nr \rightarrow inf=u_0) \wedge \circ(\neg(\beta_0 \wedge mn=nr) \rightarrow A_a)$ for every $\gamma \in \bar{\mathscr{M}}_{\psi_4}$.

Derivation of (6)

(6.1)	$\gamma \wedge A_a \rightarrow \circ B_0$ for $\gamma \not\equiv \beta_0$	(4)
(6.2)	$\gamma \wedge A_a \rightarrow \circ(\beta_0 \wedge mn=nr \rightarrow inf=u_0) \wedge$ $\circ(\neg(\beta_0 \wedge mn=nr) \rightarrow A_a)$ for $\gamma \not\equiv \beta_0$	(5), (6.1)
(6.3)	$\beta_0 \wedge mn \neq nr \wedge A_a \rightarrow \circ(mn \neq nr \wedge B_0)$	(Π)
(6.4)	$\beta_0 \wedge mn \neq nr \wedge A_a \rightarrow \circ(\beta_0 \wedge mn=nr \rightarrow inf=u_0) \wedge$ $\circ(\neg(\beta_0 \wedge mn=nr) \rightarrow A_a)$	(6.3)
(6.5)	$\beta_0 \wedge mn=nr \wedge A_a \rightarrow \circ(\beta_0 \wedge mn=nr \rightarrow inf=u_0) \wedge$ $\circ(\neg(\beta_0 \wedge mn=nr) \rightarrow A_a)$	(taut)
(6.6)	$\gamma \wedge A_a \rightarrow \circ(\beta_0 \wedge mn=nr \rightarrow inf=u_0) \wedge$ $\circ(\neg(\beta_0 \wedge mn=nr) \rightarrow A_a)$ for every $\gamma \in \bar{\mathscr{M}}_{\psi_4}$	(6.2), (6.4), (6.5)

Since $\Box \neg nil_{\psi_4}$ holds because of the absence of end labels and await statements, we get:

(7) $A_a \rightarrow inf=u_0$ **atnext** $(\beta_0 \wedge mn=nr)$

from (6) by (atnext). With

(8) $start_{\psi_4} \rightarrow A_a$

which is easy to see, we get the assertion a). For b), we let

$$A_b \equiv (\beta_0 \wedge mn=nr \rightarrow inf=u_i) \wedge (\neg(\beta_0 \wedge mn=nr) \rightarrow B_{i+1}).$$

As above, we get:

(9) $\gamma \wedge A_b \rightarrow \circ(\beta_0 \wedge mn=nr \rightarrow inf=u_{i+1}) \wedge \circ(\neg(\beta_0 \wedge mn=nr) \rightarrow A_b)$ for every $\gamma \in \bar{\mathscr{M}}_{\psi_4}$.

Derivation of (9)

(9.1)	$\gamma \wedge A_b \rightarrow \circ(\beta_0 \wedge mn=nr \rightarrow inf=u_{i+1}) \wedge$ $\circ(\neg(\beta_0 \wedge mn=nr) \rightarrow A_b)$ for $\gamma \not\equiv \beta_0$	(4), (5)
(9.2)	$\beta_0 \wedge mn \neq nr \wedge A_b \rightarrow \circ(\beta_0 \wedge mn=nr \rightarrow inf=u_{i+1}) \wedge$ $\circ(\neg(\beta_0 \wedge mn=nr) \rightarrow A_b)$	as (6.4) above
(9.3)	$\beta_0 \wedge mn=nr \wedge A_b \rightarrow nr=ls \wedge \text{nextinput}=u_{i+1}$	(2), (3)
(9.4)	$\beta_0 \wedge mn=nr \wedge A_b \rightarrow \circ(mn \neq nr \wedge nr \neq ls \wedge \text{nextinput}=u_{i+1})$	(Π), (9.3)

(9.5) $\beta_0 \wedge mn=nr \wedge A_b \to \circ(\beta_0 \wedge mn=nr \to inf=u_{i+1}) \wedge$
$\circ(\neg(\beta_0 \wedge mn=nr) \to A_b)$ (9.4)

(9.6) $\gamma \wedge A_b \to \circ(\beta_0 \wedge mn=nr \to inf=u_{i+1}) \wedge$
$\circ(\neg(\beta_0 \wedge mn=nr) \to A_b)$ for every $\gamma \in \bar{\mathcal{M}}_\Pi$ (9.1), (9.2), (9.5)

Again applying (atnext) we get:

(10) $A_b \to inf=u_{i+1}$ **atnext** $(\beta_0 \wedge mn=nr)$

and with:

(11) $\beta_0 \wedge mn=nr \wedge inf=u_i \to A_b$

which is tautologically valid, we get the assertion b). □

Our next example deals with an (idealized) implementation of a first-in-first-out queue by an infinite array *qu* and two variables *r* and *t* holding the rear and top indices, respectively:

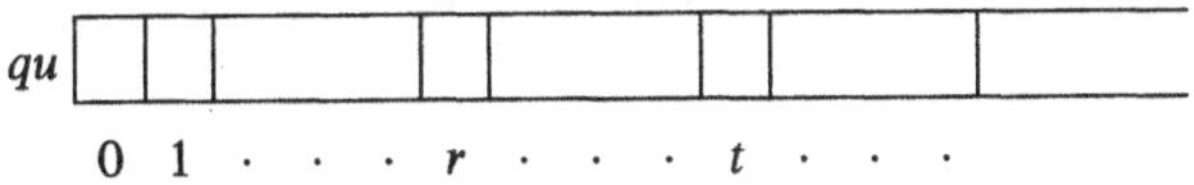

Program Example 19.2

```
ψ5 ≡ initial r=0 ∧ t=0;
     cobegin loop
                   ⋮
              α: qu[t]:=nextobj; t:=t+1;
                   ⋮
             end
           ||
             loop
                   ⋮
              β: await r≠t then a:=qu[r]; r:=r+1;
                   ⋮
             end
     coend
```

ψ_5 consists of two parallel components, the first of which continuously stores objects in the queue. The other one loads objects out of the queue into a private variable *a*, always testing whether there is an object in *qu* at all and waiting for it if this is not the case. We assume that no actions other than α and β change *qu*, *t* and *r*. Furthermore, we consider objects obj_1, obj_2, ... distinguishable by their indices. This can be expressed by:

(1) $obj_i \neq obj_k$ for every $i \neq k$.

We do not care here how the data structure "array" fits into the framework of a first-order language (in fact, this is not completely trivial). Dispensing with a complete formalization we simply view *qu*[0], *qu*[1], ... as variables in the usual sense.

We want to prove that the transfer of objects in ψ_5 is in fact performed in a first-in-first-out manner. This can be expressed by the following assertion:

Assertion

$$\psi_5 \vdash \alpha \wedge \text{nextobj} = \text{obj}_i \wedge \text{obj}_k \notin qu \rightarrow$$
$$(\beta \wedge \text{obj}_i = qu[r]) \; \mathbf{before} \; (\beta \wedge \text{obj}_k = qu[r]).$$

Here obj$\in qu$ is used as an abbreviation for $\exists i(r \leq i < t \wedge \text{obj} = qu[i])$ meaning that "object obj is contained in qu". Clearly, $\notin$ is the negation of $\in$. The assertion expresses that "if obj_i is stored in qu and obj_k is not yet contained in qu then obj_i will be got out of qu before obj_k".

Proof of the assertion. We first note that the statement α should be understood such that every "next object" nextobj is really a "new" object. This part of the specification of α is expressed by:

(2) $\alpha \rightarrow \circ(\text{nextobj} \notin qu)$

The other specification of α and the specification of β are trivial. Next we note some global invariants which are easily verified with (inv′) and (2):

(3) $\square(r \leq t)$
(4) $\square(\text{nextobj} = \text{obj}_i \rightarrow \text{obj}_i \notin qu)$
(5) $\square(qu[l] \neq qu[m])$ for $l \neq m, r \leq l, \; m < t$.

Now let

$$A_1 \equiv \text{nextobj} = \text{obj}_i \wedge \text{obj}_k \notin qu,$$
$$A_2 \equiv \text{obj}_i \in qu \wedge \text{obj}_k \notin qu,$$
$$A_3 \equiv \exists l \exists m(r \leq l < m < t \wedge \text{obj}_i = qu[l] \wedge \text{obj}_k = qu[m]),$$
$$A \equiv (\beta \rightarrow \text{obj}_i \neq qu[r]) \wedge (A_1 \vee A_2 \vee A_3).$$

We first note the simple facts:

(6) $A_1 \vee A_2 \vee A_3 \rightarrow A \vee (\beta \wedge \text{obj}_i = qu[r])$
(7) $A_1 \vee A_2 \vee A_3 \rightarrow \text{obj}_k \neq qu[r]$.

(6) is a tautology and (7) follows with (5). Now we have:

(8) $\gamma \wedge (A_1 \vee A_2 \vee A_3) \rightarrow \circ(A_1 \vee A_2 \vee A_3)$ for $\gamma \in \bar{\mathcal{M}}_{\psi_5} \setminus \{\alpha, \beta\}$

since every such γ does not change the variables in $A_1 \vee A_2 \vee A_3$, and

(9) $\alpha \wedge (A_1 \vee A_2 \vee A_3) \rightarrow \circ(A_1 \vee A_2 \vee A_3)$
(10) $\beta \wedge A \rightarrow \circ(A_1 \vee A_2 \vee A_3)$.

Derivation of (9)

(9.1)	$\alpha \wedge A_1 \rightarrow \circ A_2$	(Π), (3)
(9.2)	$\alpha \wedge A_2 \rightarrow \circ(A_2 \vee A_3)$	(Π), (4)
(9.3)	$\alpha \wedge A_3 \rightarrow \circ A_3$	(Π)
(9.4)	$\alpha \wedge (A_1 \vee A_2 \vee A_3) \rightarrow \circ(A_1 \vee A_2 \vee A_3)$	(9.1)–(9.3)

Derivation of (10)

(10.1)	$\beta \wedge A_1 \rightarrow \circ A_1$	(Π)
(10.2)	$\beta \wedge (\beta \rightarrow \text{obj}_i \neq qu[r]) \wedge A_2 \rightarrow \circ A_2$	(Π)

(10.3) $\beta \wedge (\beta \rightarrow \text{obj}_i \neq qu[r]) \wedge A_3 \rightarrow \circ A_3$ (Π)

(10.4) $\beta \wedge A \rightarrow \circ(A_1 \vee A_2 \vee A_3)$ (10.1)–(10.3)

From (6)–(10) we derive:

(11) $\gamma \wedge A \rightarrow \circ\neg(\beta \wedge \text{obj}_k = qu[r]) \wedge \circ(A \vee (\beta \wedge \text{obj}_i = qu[r]))$
for every $\gamma \in \bar{\mathscr{M}}_{\psi_5}$.

Trivially we have:

(12) $\text{nil}_{\psi_5} \wedge A \rightarrow \neg(\beta \wedge \text{obj}_k = qu[r])$

and applying the rule (before) we get:

(13) $A \rightarrow (\beta \wedge \text{obj}_i = qu[r])$ **before** $(\beta \wedge \text{obj}_k = qu[r])$.

With

(14) $\alpha \wedge \text{nextobj} = \text{obj}_i \wedge \text{obj}_k \notin qu \rightarrow A$

which is tautologically valid, the assertion follows immediately. □

Our third example is another solution of the mutual exclusion problem discussed in Section 16.

Program Example 19.3

```
ψ6 ≡ initial a1=false ∧ a2=false ∧ t=1;
   cobegin loop
                    ⋮
              α1: a1:=true;
              α2: t:=1;
              α3: await a2=false ∨ t=2;
              α4: ~ ⎫
                  ⋮ ⎭ cs1
              α5: a1:=false;
                    ⋮
           end
         ‖
           loop
                    ⋮
              β1: a2:=true;
              β2: t:=2;
              β3: await a1=false ∨ t=1;
              β4: ~ ⎫
                  ⋮ ⎭ cs2
              β5: a2:=false;
                    ⋮
           end
   coend
```

(No statements other than the displayed ones change the variables a_1, a_2 and t.)

This solution (due to Peterson (1981)) needs no semaphore operations. We do not want to show the correct mutual exclusion of cs_1 and cs_2 but the fact that "if one of the parallel components wants to enter its critical section it may, at the latest, do this after the other component has entered its critical section once" (1-bounded overtaking). Since the two components are symmetric it is sufficient to formulate this claim for, say, Π_1:

Assertion

$$\psi_6 \vdash \text{at } \alpha_3 \rightarrow \text{at } \alpha_4 \ \mathbf{atnext}\ (\text{at } \alpha_4 \vee \text{at } \beta_4) \vee \text{at } \alpha_4 \ \mathbf{atnext}^2\ (\text{at } \alpha_4 \vee \text{at } \beta_4).$$

Proof of the assertion

Let

L_1 = set of labels between α_2 and α_5, both included,
L_2 = set of labels between β_2 and β_5, both included.

We again first note some quite trivial invariants:

(1) $\square(t=1 \vee t=2)$
(2) $\square(a_1 = \mathbf{true} \leftrightarrow \text{at } L_1)$
(3) $\square(a_2 = \mathbf{true} \leftrightarrow \text{at } L_2)$
(4) $\square \neg \text{nil}_{\psi_6}$.

Then we let

$A \equiv \text{at } \alpha_3 \wedge \text{at } \beta_3 \wedge t=1$
$B_1 \equiv \text{at } \alpha_3 \wedge (\text{at } \beta_3 \rightarrow t=2)$

and get

(5) $\alpha \wedge A \rightarrow \circ(\text{at } \alpha_4 \vee \text{at } \beta_4 \rightarrow B_1) \wedge \circ(\neg(\text{at } \alpha_4 \vee \text{at } \beta_4) \rightarrow A)$
for every $\alpha \in \bar{\mathcal{M}}_{\psi_6}$
(6) $\alpha \wedge B_1 \rightarrow \circ(\text{at } \alpha_4 \vee \text{at } \beta_4 \rightarrow \text{at } \alpha_4) \wedge \circ(\neg(\text{at } \alpha_4 \vee \text{at } \beta_4) \rightarrow B_1)$
for every $\alpha \in \bar{\mathcal{M}}_{\psi_6}$.

Derivation of (5). Let $C \equiv \circ(\text{at } \alpha_4 \vee \text{at } \beta_4 \rightarrow B_1) \wedge \circ(\neg(\text{at } \alpha_4 \vee \text{at } \beta_4) \rightarrow A)$.

(5.1)	$\alpha_3 \wedge A \rightarrow a_2 = \mathbf{true} \wedge t=1 \wedge (a_2 = \mathbf{false} \vee t=2)$	(3), (Π)
(5.2)	$\alpha_3 \wedge A \rightarrow C$	(5.1)
(5.3)	$\beta_3 \wedge A \rightarrow \circ(\text{at } \alpha_3 \wedge \text{at } \beta_4)$	(Π)
(5.4)	$\text{at } \alpha_3 \wedge \text{at } \beta_4 \rightarrow B_1$	(taut)
(5.5)	$\beta_3 \wedge A \rightarrow C$	(5.3), (5.4)
(5.6)	$\gamma \wedge A \rightarrow C$ for $\gamma \in \bar{\mathcal{M}}_{\psi_6} \setminus \{\alpha_3, \beta_3\}$	(Π)
(5.7)	$\alpha \wedge A \rightarrow C$ for every $\alpha \in \bar{\mathcal{M}}_{\psi_6}$	(5.2), (5.5), (5.6)

The derivation of (6) runs analogously and is left as an exercise. From (5) and (6), together with (4), we get:

(7) $A \vee B_1 \rightarrow \text{at } \alpha_4 \ \mathbf{atnext}\ (\text{at } \alpha_4 \vee \text{at } \beta_4) \vee \text{at } \alpha_4 \ \mathbf{atnext}^2 (\text{at } \alpha_4 \vee \text{at } \beta_4)$

by the rule (disatnext^2), and with

(8) at $\alpha_3 \rightarrow A \vee B_1$

which follows immediately from (1), we get the assertion.

Observe that this proof also shows in which case an overtaking really happens. It is the case described by the formula A (which yields the **atnext**^{2} part), i.e., the case in which Π_2 also wants to enter its critical section and t has the value 1. In the complementary case (Π_2 not at β_3 or $t \neq 2$), described by B_1, we have $B_1 \rightarrow$ at α_4 **atnext** (at $\alpha_4 \vee$ at β_4) by (atnext) and no overtaking takes place. □

Chapter VI
Eventuality Properties of Programs

20. Fair Execution Sequences

We now want to investigate the verification of eventuality properties of programs expressing that "eventually something will happen". However, we first have to come back to our interleaving model of computation and the notion of execution sequences of programs and have a somewhat closer look at them.

Consider the sample program:

$$\begin{array}{l} \Pi \equiv \textbf{initial } a=1; \\ \quad \textbf{cobegin } \alpha_0: a:=0; \\ \qquad\qquad \alpha_1: \textbf{stop} \\ \qquad\qquad \| \\ \qquad\qquad \beta_0: \textbf{while } a=1 \textbf{ do } \beta_1: b:=0 \textbf{ od}; \\ \qquad\qquad \beta_2: \textbf{stop} \\ \quad \textbf{coend} \end{array}$$

In any reasonable idea of really parallel execution of Π (on two processors) we would clearly claim that Π terminates sometime or, in particular, β_2 is eventually reached. This is because execution of α_0 should be carried out sometime and will set a to 0 so that the other parallel component will leave its loop at the next test $a=1$.

Thus, the formula:

$$\text{start}_\Pi \rightarrow \Diamond \text{ at } \beta_2$$

should be Π-valid, but unfortunately it is not. Remember that Π-validity means validity "in every execution sequence" of Π. Now there is one execution sequence of Π which is caused by a "very unfair scheduling" and for which $\text{start}_\Pi \rightarrow \Diamond$ at β_2 is obviously not true. This is the sequence where in every state the second component of Π is scheduled so that α_0 is never executed and β_0 and β_1 are executed in turn forever:

$$\beta_0; \beta_1; \beta_0; \beta_1; \ldots$$

This example shows in fact an adulteration of the intuitive idea of parallelism by interleaving. Note that such strange execution sequences do not cause any harm with invariance and precedence properties but only with eventuality properties.

In order to overcome this insufficiency of the interleaving model we supply it with an additional condition which prohibits execution sequences such as those above. In terms of the model we may express such an appropriate *fair scheduling assumption* as follows:

> "Any statement which may be scheduled infinitely often, will be executed infinitely often".

With respect to the formal definition of execution sequences (cf. Section 13) this assumption gives rise to the following definition.

Definition. An execution sequence $\mathbf{W}_\Pi = \{\eta_0, \eta_1, \eta_2, \ldots\}$ of a program Π (w.r.t. some S, ξ) is called *fair* if it has the following property: if there are infinitely many states $\eta_k = (\mu_k, \lambda_k^{(1)}, \ldots, \lambda^{(i)}, \ldots, \lambda_k^{(p)}, \kappa_k)$ in $\mathbf{W}_\Pi$ containing the same $\lambda^{(i)}$ such that $\mathbf{S}^{(\xi,\mu_k)}(E_{\lambda^{(i)}}) = \mathbf{t}$ then $\kappa_k = i$ in infinitely many η_k.

In fact, this definition does not contain the execution sequence involved by the unfair scheduling discussed above since this sequence is of the form:

$$\begin{aligned}
\eta_0 &= (\mu_0, \alpha_0, \beta_0, 2),\\
\eta_1 &= (\mu_1, \alpha_0, \beta_1, 2),\\
\eta_2 &= (\mu_2, \alpha_0, \beta_0, 2),\\
\eta_3 &= (\mu_3, \alpha_0, \beta_1, 2),\\
\eta_4 &= (\mu_4, \alpha_0, \beta_0, 2),\\
&\vdots
\end{aligned}$$

α_0 occurs in every η_k, $E_{\alpha_0} = \mathbf{true}$, hence $\mathbf{S}^{(\xi,\mu_k)}(E_{\alpha_0}) = \mathbf{t}$ for all S and ξ, but it is not $\kappa = 1$ in infinitely many η_k (there is even $\kappa = 2$ in every η_k). On the other hand, every interleaving which schedules α_0 sometime yields a fair execution sequence, e.g.,

$$\begin{aligned}
\eta_0 &= (\mu_0, \alpha_0, \beta_0, 2),\\
\eta_1 &= (\mu_1, \alpha_0, \beta_1, 2),\\
\eta_2 &= (\mu_2, \alpha_0, \beta_0, 2),\\
\eta_3 &= (\mu_3, \alpha_0, \beta_1, 1),\\
\eta_4 &= (\mu_4, \alpha_1, \beta_1, 2),\\
\eta_5 &= (\mu_4, \alpha_1, \beta_0, 2),\\
\eta_6 &= (\mu_4, \alpha_1, \beta_2, 0),\\
&\vdots
\end{aligned}$$

In any such case there are only finitely many η_k containing α_0.

From now on we want to consider formulas which are Π-valid under the additional assumption that only fair execution sequences are permitted as state sequences in temporal structures. Axiomatically we can describe this restriction by an additional basic program axiom (λ ranging over $\bar{\mathcal{M}}_\Pi$) which is easily shown to be sound with respect to the above definition:

(B3) $\Box\Diamond(\text{at } \lambda \wedge E_\lambda) \rightarrow \Diamond\lambda$
"If λ is enabled infinitely often then λ is executed sometime".

Note that with (T41) we immediately get:

$$\Box\Diamond(\text{at } \lambda \wedge E_\lambda) \rightarrow \Box\Diamond\lambda$$

from (B3) which makes the coincidence with the semantic definition even more evident.

We want briefly to indicate that this fair scheduling assumption is not the only reasonable possibility of restricting to "fairness". Let us first note the following modified assumption:

> "Any parallel component which may be scheduled infinitely often, will be scheduled infinitely often".

In order to give an axiomatic description of this assumption we let $\bar{\mathscr{M}}_{\Pi_i} = \{\alpha_0^{(i)}, \ldots, \alpha_m^{(i)}\}$ and

$$E^{(i)} \equiv (\text{at } \alpha_0^{(i)} \wedge E_{\alpha_0^{(i)}}) \vee \ldots \vee (\text{at } \alpha_m^{(i)} \wedge E_{\alpha_m^{(i)}})$$

("one of the actions of Π_i is enabled").

With this notation we can give the program axiom:

(B3′) $\quad \Box\Diamond E^{(i)} \rightarrow \Diamond(\alpha_0^{(i)} \vee \ldots \vee \alpha_m^{(i)}) \quad$ (for every $i = 1, \ldots, p$)

which provides the desired restriction. (B3′) seems to be more general than (B3). However, it turns out that both are equivalent: we can derive (B3′) from (B3) and vice versa.

Before we prove this we remember a logical law derived in Section 6:

(T47) $\quad \Box\Diamond(A \vee B) \leftrightarrow \Box\Diamond A \vee \Box\Diamond B.$

Derivation of (B3)⊢(B3′)

(1)	$\Box\Diamond(\text{at } \lambda \wedge E_\lambda) \rightarrow \Diamond\lambda$	assumption
(2)	$\Box\Diamond E^{(i)} \rightarrow \Box\Diamond(\text{at } \alpha_0^{(i)} \wedge E_{\alpha_0^{(i)}}) \vee \ldots \vee \Box\Diamond(\text{at } \alpha_m^{(i)} \wedge E_{\alpha_m^{(i)}})$	(T47)
(3)	$\Box\Diamond(\text{at } \alpha_j^{(i)} \wedge E_{\alpha_j^{(i)}}) \rightarrow \Diamond\alpha_j^{(i)}$ for every $j = 0, \ldots, m$	(1)
(4)	$\Box\Diamond E^{(i)} \rightarrow \Diamond(\alpha_0^{(i)} \vee \ldots \vee \alpha_m^{(i)})$	(2), (3) □

Derivation of (B3′)⊢(B3). We assume $\lambda \equiv \alpha_j^{(i)}$ for some appropriate i and j.

(1)	$\Box\Diamond E^{(i)} \rightarrow \Diamond(\alpha_0^{(i)} \vee \ldots \vee \alpha_m^{(i)})$	assumption
(2)	$\Box\Diamond(\text{at } \lambda \wedge E_\lambda) \wedge \Box\Box\neg\lambda \rightarrow \Diamond(\text{at } \lambda \wedge E_\lambda \wedge \Box\neg\lambda)$	(T39), (T4)
(3)	$\text{at } \lambda \wedge \Box\neg\lambda \rightarrow \circ(\text{at } \lambda \wedge \Box\neg\lambda)$	(Π5)
(4)	$\text{at } \lambda \wedge \Box\neg\lambda \rightarrow \Box(\text{at } \lambda \wedge \Box\neg\lambda)$	(ind′), (3)
(5)	$\text{at } \lambda \wedge \neg\lambda \rightarrow \neg(\alpha_0^{(i)} \vee \ldots \vee \alpha_m^{(i)})$	(Π2)
(6)	$\Box\Diamond(\text{at } \lambda \wedge E_\lambda) \wedge \Box\neg\lambda \rightarrow \Diamond\Box\neg(\alpha_0^{(i)} \vee \ldots \vee \alpha_m^{(i)})$	(2), (4), (5)
(7)	$\Box\Diamond(\text{at } \lambda \wedge E_\lambda) \rightarrow \Box\Diamond E^{(i)}$	definition of $E^{(i)}$
(8)	$\Box\Diamond(\text{at } \lambda \wedge E_\lambda) \rightarrow \Box\Diamond(\alpha_0^{(i)} \vee \ldots \vee \alpha_m^{(i)})$	(1), (T41)
(9)	$\Box\Diamond(\text{at } \lambda \wedge E_\lambda) \rightarrow \Diamond\lambda$	(6), (8) □

Another, actually weaker, kind of fairness is given by the following *justice assumption*:

> "If some statement may be scheduled permanently from some point on then it will be executed infinitely often",

or, again, in a somewhat modified version:

> "If some parallel component may be scheduled permanently from some point on then it will be scheduled infinitely often".

Axiomatically these restrictions are described by:

(J) $\Diamond\Box(\text{at }\lambda \wedge E_\lambda) \rightarrow \Diamond \lambda$,

(J′) $\Diamond\Box E^{(i)} \rightarrow \Diamond(\alpha_0^{(i)} \vee \ldots \vee \alpha_m^{(i)})$ (for every $i=1, \ldots, p$),

respectively. (J) and (J′) can again be proved to be equivalent.

It is almost trivial to see that every fair execution sequence is *just*, i.e., that (B3)⊢(J) holds:

Derivation

(1)	$\Box\Diamond(\text{at }\lambda \wedge E_\lambda) \rightarrow \Diamond \lambda$	assumption
(2)	$\Diamond\Box(\text{at }\lambda \wedge E_\lambda) \rightarrow \Box\Diamond(\text{at }\lambda \wedge E_\lambda)$	(T10)
(3)	$\Diamond\Box(\text{at }\lambda \wedge E_\lambda) \rightarrow \Diamond \lambda$	(1), (2) □

Furthermore we note that justice and fairness are equivalent in the absence of await statements. In this case every E_λ is identically **true**, thus (B3) and (J) reduce to $\Box\Diamond \text{ at }\lambda \rightarrow \Diamond\lambda$ and $\Diamond\Box \text{ at }\lambda \rightarrow \Diamond\lambda$, respectively. As in the derivation of (B3′)⊢(B3) we similarly find $\Box\Diamond \text{ at }\lambda \wedge \Box\neg\lambda \rightarrow \Diamond\Box \text{ at}\lambda$ and applying (J) we get (B3).

As a last example we consider the following *impartiality assumption*, which only makes sense in cases when none of the parallel components terminates:

"Every parallel component is scheduled infinitely often".

It is axiomatized by:

(Imp) $\Diamond(\alpha_0^{(i)} \vee \ldots \vee \alpha_m^{(i)})$ (for every $i=1, \ldots, p$)

and is clearly stronger than fairness since (B3′) is trivially derived from (Imp). For cyclic programs without await statements impartiality and fairness are again equivalent since then (B3′) reduces to

$$\Box\Diamond(\text{at }\alpha_0^{(i)} \vee \ldots \vee \text{at }\alpha_m^{(i)}) \rightarrow \Diamond(\alpha_0^{(i)} \vee \ldots \vee \alpha_m^{(i)})$$

and $\text{at}\alpha_0^{(i)} \vee \ldots \vee \text{at}\alpha_m^{(i)}$ can be shown to be identically **true** ("in any state, Π_i must be located at some label $\alpha_0^{(i)}, \ldots, \alpha_m^{(i)}$"). So (B3′) reads $\Box\Diamond$ **true** $\rightarrow \Diamond(\alpha_0^{(i)} \vee \ldots \vee \alpha_m^{(i)})$, and from this (Imp) follows trivially since $\Box\Diamond$ **true** is valid.

In the subsequent sections we do not follow up these various concepts. We always assume fairness given by the basic axiom (B3) (or its equivalent form (B3′)).

21. The Finite Chain Reasoning Method

Let

$\Pi \equiv$ **initial** R;
  cobegin $\Pi_1 \parallel \ldots \parallel \Pi_p$ **coend**

be a program. An eventuality property of Π is expressed by some formula:

$$A \rightarrow \Diamond B$$

and we remember the logical means for proving such formulas, discussed in Section 10. A simple first method is given by the rule:

(chain) $A \to \Diamond B,\ B \to \Diamond C \vdash A \to \Diamond C.$

In order to prove $A \to \Diamond B$ we have to prove two or – in general – finitely many "smaller steps" of the same kind. In fact there are many examples where we can proceed according to this *finite chain reasoning method.* We only have to provide a reasonable rule for providing the "smaller steps". In Section 10 we noted the simple rule:

(som) $A \to \circ B \vdash A \to \Diamond B$

for this purpose. Using (trans) this rule could be translated to:

$$\begin{array}{ll} \alpha \wedge A \to \circ B & \text{for every } \alpha \in \bar{\mathcal{M}}_\Pi \\ \mathrm{nil}_\Pi \wedge A \to B & \\ \vdash A \to \Diamond B. & \end{array}$$

However, this rule has the rather strong premise that *every* $\alpha \in \bar{\mathcal{M}}_\Pi$ leads from A to B. We want to formulate a more useful rule where we require that *some* α (which cannot be delayed forever) leads from A to B while all other actions leave A invariant:

$$\begin{array}{ll} \text{(event0)} & A \ \mathtt{invof}\ \bar{\mathcal{M}}_\Pi \backslash \{\alpha\}, \\ & \alpha \wedge A \to \circ B, \\ & \Box A \to \Diamond(\mathrm{at}\ \alpha \wedge E_\alpha) \\ & \quad (\text{for arbitrary } \alpha \in \bar{\mathcal{M}}_\Pi) \\ & \vdash A \to \Diamond B. \end{array}$$

The third premise in this rule guarantees that α is executed sometime, since otherwise $\Box A$ would hold and hence we would have $\Diamond(\mathrm{at}\,\alpha \wedge E_\alpha)$ and also $\Box\Diamond(\mathrm{at}\,\alpha \wedge E_\alpha)$. By the fair scheduling assumption, α will eventually be executed. We give a formal proof in the following:

Theorem 21.1. *The rule (event0) is Π-derived.*

Proof. We give a direct derivation.

(1)	$A\ \mathtt{invof}\ \bar{\mathcal{M}}_\Pi \backslash \{\alpha\}$	assumption
(2)	$\alpha \wedge A \to \circ B$	assumption
(3)	$\Box A \to \Diamond(\mathrm{at}\ \alpha \wedge E_\alpha)$	assumption
(4)	$\gamma \wedge \neg\alpha \wedge A \to \circ A$ for every $\gamma \in \bar{\mathcal{M}}_\Pi \backslash \{\alpha\}$	(1)
(5)	$\alpha \wedge \neg\alpha \wedge A \to \circ A$	(taut)
(6)	$\mathrm{nil}_\Pi \wedge \neg\alpha \wedge A \to A$	(taut)
(7)	$\neg\alpha \wedge A \to \circ A$	(trans), (4), (5), (6)
(8)	$\Box\neg(\alpha \wedge A) \to \neg(\alpha \wedge A) \wedge \circ\Box\neg(\alpha \wedge A)$	(ax 3)
(9)	$A \wedge \Box\neg(\alpha \wedge A) \to A \wedge \neg\alpha \wedge \circ\Box\neg(\alpha \wedge A)$	(8)
(10)	$A \wedge \Box\neg(\alpha \wedge A) \to \circ(A \wedge \Box\neg(\alpha \wedge A))$	(7), (9)
(11)	$A \wedge \Box\neg(\alpha \wedge A) \to \Box(A \wedge \Box\neg(\alpha \wedge A))$	(ind′), (10)

(12)	$A \wedge \Box\neg(\alpha \wedge A) \rightarrow \Box A \wedge \Box\neg\alpha$	(11)
(13)	$\Box A \rightarrow \Box\Diamond(\text{at } \alpha \wedge E_\alpha)$	(T41), (3)
(14)	$\Box\Diamond(\text{at } \alpha \wedge E_\alpha) \rightarrow \Diamond\alpha$	(B3)
(15)	$A \wedge \Box\neg(\alpha \wedge A) \rightarrow \Diamond\alpha \wedge \Box\neg\alpha$	(12), (13), (14)
(16)	$A \rightarrow \neg\Box\neg(\alpha \wedge A)$	(15)
(17)	$A \rightarrow \Diamond \circ B$	(2), (16)
(18)	$A \rightarrow \Diamond B$	(17) □

Observe, once more, that the proof of this rule uses only the basic program axioms, here (B2) and (B3).

(event0) is a quite general rule. We still note some special cases which frequently occur in applications. First consider the case where the formula A is of the form:

$$\text{at } \beta_i \wedge A'$$

for some $\beta_i \in \bar{\mathcal{M}}_{\Pi_i}$, $i \in \{1, \ldots, p\}$, and α is β_i. Writing again A for A' we get:

(event1) $\alpha \wedge \text{at } \beta_i \wedge A \rightarrow \circ A$ for every $\alpha \in \bar{\mathcal{M}}_\Pi \setminus \bar{\mathcal{M}}_{\Pi_i}$,
$\beta_i \wedge A \rightarrow \circ B$,
$\Box(\text{at } \beta_i \wedge A) \rightarrow \Diamond E_{\beta_i}$
$\vdash \text{at } \beta_i \wedge A \rightarrow \Diamond B$.

Derivation of (event1)

(1)	$\alpha \wedge \text{at } \beta_i \wedge A \rightarrow \circ A$ for every $\alpha \in \bar{\mathcal{M}}_\Pi \setminus \bar{\mathcal{M}}_{\Pi_i}$	assumption
(2)	$\beta_i \wedge A \rightarrow \circ B$	assumption
(3)	$\Box(\text{at } \beta_i \wedge A) \rightarrow \Diamond E_{\beta_i}$	assumption
(4)	$\alpha \rightarrow \text{at } \alpha$	(Π2)
(5)	$\alpha \rightarrow \neg \text{ at } \beta_i$ for $\alpha \in \bar{\mathcal{M}}_{\Pi_i}, \alpha \not\equiv \beta_i$	(Π3), (4)
(6)	$\alpha \wedge \text{at } \beta_i \wedge A \rightarrow \circ(\text{at } \beta_i \wedge A)$	(5)
(7)	$(\text{at } \beta_i \wedge A)$ **invof** $\bar{\mathcal{M}}_\Pi \setminus \{\beta_i\}$	(1), (6), (Π5)
(8)	$\beta_i \wedge \text{at } \beta_i \wedge A \rightarrow \circ B$	(2)
(9)	$\Box(\text{at } \beta_i \wedge A) \rightarrow \Box \text{ at } \beta_i \wedge \Diamond E_{\beta_i}$	(3)
(10)	$\Box(\text{at } \beta_i \wedge A) \rightarrow \Diamond(\text{at } \beta_i \wedge E_{\beta_i})$	(T39), (9)
(11)	$\text{at } \beta_i \wedge A \rightarrow \Diamond B$	(event0), (7), (8), (10) □

This time we have also used some structural program axioms in deriving (event1).

A particularly simple special case is that with $A \equiv$ **true** in (event1):

(event2) $\beta_i \rightarrow \circ B$, $\Box \text{ at } \beta_i \rightarrow \Diamond E_{\beta_i} \vdash \text{at } \beta_i \rightarrow \Diamond B$.

Its proof is obvious since the first premise in (event1) is trivially fulfilled.

Further special cases are obtained if α or β_i, respectively, in (event0), (event1), or (event2) are not of the form **await** B .... Then E_α (E_{β_i}) is identically **true** and in any case the last premise holds trivially. This is also the only case where we really have reduced the problem of proving some $A \rightarrow \Diamond B$ to the proof of other formulas. Otherwise, the last premises in these rules are still eventuality properties themselves. In applications we will infer them by (som) or the likewise trivial rule:

$$A \rightarrow B \vdash A \rightarrow \Diamond B$$

immediately following from (T 32).

As a first example of such an application, let us take the program Π from the previous sections:

```
Π ≡ initial a=1;
    cobegin α0: a:=0;
            α1: stop
            ||
            β0: while a=1 do β1: b:=0 od;
            β2: stop
    coend
```

As is desirable, we are now able to prove the termination of the while loop:

Derivation of $\text{start}_\Pi \rightarrow \Diamond$ at β_2. Let $A \equiv (\text{at } \beta_0 \vee \text{at } \beta_1) \wedge a=1$. We assert the following eventualities:

(1) $\text{at } \alpha_0 \wedge A \rightarrow \Diamond((\text{at } \beta_0 \vee \text{at } \beta_1) \wedge a=0)$
(2) $\text{at } \beta_0 \wedge a=0 \rightarrow \Diamond \text{ at } \beta_2$
(3) $\text{at } \beta_1 \wedge a=0 \rightarrow \Diamond(\text{at } \beta_0 \wedge a=0)$.

Each of these formulas is derived by (event1), the last premise being trivial in every case. The other respective premises are listed below:

(1.1) $\beta_0 \wedge \text{at } \alpha_0 \wedge A \rightarrow \circ A$ (Π)
(1.2) $\beta_1 \wedge \text{at } \alpha_0 \wedge A \rightarrow \circ A$ (Π)
(1.3) $\alpha_0 \wedge A \rightarrow \circ((\text{at } \beta_0 \vee \text{at } \beta_1) \wedge a=0)$ (Π)

(2.1) $\alpha_0 \wedge \text{at } \beta_0 \wedge a=0 \rightarrow \circ(a=0)$ (Π)
(2.2) $\beta_0 \wedge a=0 \rightarrow \circ \text{ at } \beta_2$ (Π)

(3.1) $\alpha_0 \wedge \text{at } \beta_1 \wedge a=0 \rightarrow \circ(a=0)$ (Π)
(3.2) $\beta_1 \wedge a=0 \rightarrow \circ(\text{at } \beta_0 \wedge a=0)$ (Π)

Now from (2) and (3) we get:

(4) $\text{at } \beta_1 \wedge a=0 \rightarrow \Diamond \text{ at } \beta_2$

by (chain), and applying this rule again to (1), (2), (4) we get:

(5) $\text{at } \alpha_0 \wedge A \rightarrow \Diamond \text{ at } \beta_2$.

With the trivial

(6) $\text{start}_\Pi \rightarrow \text{at } \alpha_0 \wedge A$,

we then get the desired formula:

(7) $\text{start}_\Pi \rightarrow \Diamond \text{ at } \beta_2$. □

22. The Method of Well-Founded Orderings

We have already argued in Section 10 that the chain reasoning method expressed in (chain) is not universally applicable. Let us repeat the rather abstract argumentation there in the concrete context of a simple program verification problem. Consider the sequential program:

$\Pi \equiv$ **initial** $a \geq 0$;
  α_0: **while** $a>0$ **do** α_1: $a := a-1$ **od**;
  α_2: **stop**

It is intuitively clear that Π terminates, i.e.,

(1) $\quad \text{at } \alpha_0 \wedge a \geq 0 \rightarrow \Diamond \text{ at } \alpha_2.$

If we want to prove (1) with (chain) we may first try to find a "shorter step":

(2) $\quad \text{at } \alpha_0 \wedge a \geq 0 \rightarrow \Diamond B$

provable by one of the other rules discussed in the previous section. In fact, we have not much choice since the most general (event0) reduces in this case to the simple (som):

$$\alpha_0 \wedge a \geq 0 \rightarrow \circ B \vdash \text{at } \alpha_0 \wedge a \geq 0 \rightarrow \Diamond B.$$

Since $\alpha_0 \wedge a \geq 0 \rightarrow \circ(\text{at } \alpha_2 \vee (\text{at } \alpha_1 \wedge a>0))$ is the "best possible" premise, we get:

(3) $\quad \text{at } \alpha_0 \wedge a \geq 0 \rightarrow \Diamond(\text{at } \alpha_2 \vee (\text{at } \alpha_1 \wedge a>0))$

as the "best" instance of (2). Now we have to follow up the alternative $\text{at}\,\alpha_1 \wedge a>0$ in (3). The "best" we can imply from this formula is $\Diamond(\text{at } \alpha_0 \wedge a \geq 0)$:

(4) $\quad \text{at } \alpha_1 \wedge a>0 \rightarrow \Diamond(\text{at } \alpha_0 \wedge a \geq 0).$

Obviously we now are in an infinite "derivation loop" since the next chain we would have to prove according to (3) and (4) is $\text{at}\,\alpha_0 \wedge a \geq 0 \rightarrow \Diamond \text{ at } \alpha_2$, which is just our starting point (1).

The problem with this trial is that the immediately convincing informal argument that the variable a properly decreases with every loop run and hence cannot be >0 forever does not appear. In order to mirror this idea we note that to prove (1) it is sufficient to show

(5) $\quad \text{at } \alpha_0 \wedge a=n \rightarrow \Diamond \text{ at } \alpha_2$

where n is a variable ranging over $\mathbb{N}_0$. In fact, from (5) we deduce $\exists n(\text{at}\,\alpha_0 \wedge a=n) \rightarrow \Diamond\,\text{at}\,\alpha_2$ with (pred) and because of $\text{at}\,\alpha_0 \wedge a \geq 0 \rightarrow \exists n(\text{at } \alpha_0 \wedge a=n)$ we then get (1). Similarly to the earlier case, we now get:

(6) $\quad \text{at } \alpha_0 \wedge a=n \rightarrow \Diamond(\text{at } \alpha_2 \vee (\text{at } \alpha_0 \wedge a=n-1))$

and have avoided the above circle in reducing $\text{at}\,\alpha_0 \wedge a=n$ to $\text{at}\,\alpha_0 \wedge a=n-1$. By applying this argument n times (i.e., by induction on n) we find $\text{at}\,\alpha_0 \wedge a=n \rightarrow \Diamond(\text{at } \alpha_2 \vee (\text{at } \alpha_0 \wedge a=0))$ which provides (5) because of $\text{at}\,\alpha_0 \wedge a=0 \rightarrow \Diamond\text{at } \alpha_2$.

This latter argumentation is just an example of the idea which led us to the general principle of well-founded orderings in Section 10 expressed by the rule:

(wfo) $A(z) \to \Diamond (B \vee \exists z'(z' \prec z \wedge A(z'))) \vdash \exists z A(z) \to \Diamond B$

where we assume the underlying logical language to be some $\mathscr{L}_{\mathrm{TP}}^{\mathrm{wf}}$, z and z' are global variables over a well-founded set and B does not contain z.

This rule is directly applicable in the case of sequential programs. We illustrate this by a somewhat less trivial example.

Example. Let

$\Pi \equiv$ **initial** $a=0 \wedge b=1 \wedge c=1 \wedge n \geq 0;$
α_0 : **while** $c \leq n$ **do** α_1: $a := a+1;$
α_2: $b := b+2;$
α_3: $c := c+b$ **od**;
α_4 : **stop**

Π computes entier$(\sqrt{n})$ but here we only want to show termination, i.e.,

$$\mathrm{start}_\Pi \to \Diamond \text{ at } \alpha_4.$$

For the proof we let:

$$Z = \{z = (z_1, z_2) \in \mathbf{N}_0^2 \mid z_1 > 0 \text{ and } z_2 - z_1 \leq n\},$$
$$(z_1, z_2) \preccurlyeq (z_1', z_2') \text{ iff } (z_1, z_2) = (z_1', z_2') \text{ or } z_2 - z_1 > z_2' - z_1'.$$

$\preccurlyeq$ can easily be seen to be a well-founded ordering on Z. With $A(z) \equiv \mathrm{at}\, \alpha_0 \wedge b = z_1 \wedge c = z_2$ we then have:

(1) $A(z) \to \circ(\mathrm{at}\ \alpha_4 \vee (\mathrm{at}\ \alpha_1 \wedge b = z_1 \wedge c = z_2 \wedge c \leq n))$ (Π)
(2) $\mathrm{at}\ \alpha_1 \wedge b = z_1 \wedge c = z_2 \wedge c \leq n \to$
$\Diamond(\mathrm{at}\ \alpha_0 \wedge b = z_1 + 2 \wedge c = z_2 + z_1 + 2 \wedge z_2 \leq n)$ (Π)
(3) $\mathrm{at}\ \alpha_0 \wedge b = z_1 + 2 \wedge c = z_2 + z_1 + 2 \wedge z_2 \leq n \to \exists z'(z' \prec z \wedge A(z'))$ (data)
(4) $A(z) \to \Diamond(\mathrm{at}\ \alpha_4 \vee \exists z'(z' \prec z \wedge A(z')))$ (1), (2), (3)
(5) $\exists z A(z) \to \Diamond\ \mathrm{at}\ \alpha_4$ (wfo), (4)
(6) $\mathrm{start}_\Pi \to \exists z A(z)$ (pred)
(7) $\mathrm{start}_\Pi \to \Diamond\ \mathrm{at}\ \alpha_4$ (5), (6) □

In the case of parallel programs the rule (wfo) is usually not applicable in this direct way. The reason is that not every single action may be expected to decrease the "parameter" z in $A(z)$. However, it would suffice if we could prove that:

- at least some "helpful" statements decrease z in $A(z)$,
- every other statement does not increase z in $A(z)$,
- one of the helpful statements is eventually enabled.

We are quite free to choose any helpful set of statements and even this need not be done once and for all in some program proof, but may depend on the values of any variables.

We give a standard form of this proof method in which the helpful set of statements may depend on z and is in every case just the set of statements of some parallel component Π_i. Which component is chosen for a particular z is described by a *helpfulness function*:

$$h\colon\ Z \to \{1, \ldots, p\}$$

which determines the index i of Π_i. Recalling the notation $E^{(i)} \equiv (\text{at } \alpha_0^{(i)} \wedge E_{\alpha_0^{(i)}}) \vee \ldots \vee (\text{at } \alpha_m^{(i)} \wedge E_{\alpha_m^{(i)}})$ where $\bar{\mathcal{M}}_{\Pi_i} = \{\alpha_0^{(i)}, \ldots, \alpha_m^{(i)}\}$ we formulate the rule:

(well)	$\alpha \wedge A(z) \to \circ(B \vee \exists z'(z' \preccurlyeq z \wedge A(z')))$	for every $\alpha \in \bar{\mathcal{M}}_\Pi \setminus \bar{\mathcal{M}}_{\Pi_{h(z)}}$,
	$\alpha \wedge A(z) \to \circ(B \vee \exists z'(z' \prec z \wedge A(z')))$	for every $\alpha \in \bar{\mathcal{M}}_{\Pi_{h(z)}}$,
	$\Box A(z) \to \Diamond(B \vee E^{(h(z))})$	
	$\vdash \exists z A(z) \to \Diamond B$	
		(z not contained in B)

Theorem 22.1. *The rule* (well) *is* Π*-derived.*

Proof. Let $\bar{\mathcal{M}}_{\Pi_{h(z)}} = \{\beta_1, \ldots, \beta_m\}$. We give a derivation of (well):

(1)	$\alpha \wedge A(z) \to \circ(B \vee \exists z'(z' \preccurlyeq z \wedge A(z')))$ for every $\alpha \in \bar{\mathcal{M}}_\Pi \setminus \bar{\mathcal{M}}_{\Pi_{h(z)}}$	assumption
(2)	$\alpha \wedge A(z) \to \circ(B \vee \exists z'(z' \prec z \wedge A(z')))$ for every $\alpha \in \bar{\mathcal{M}}_{\Pi_{h(z)}}$	assumption
(3)	$\Box A(z) \to \Diamond(B \vee E^{(h(z))})$	assumption
(4)	$A(z) \to \exists z'(z' \preccurlyeq z \wedge A(z'))$	(pred)
(5)	$\alpha \wedge \exists z'(z' \preccurlyeq z \wedge A(z')) \wedge \Box\neg B \to \circ(\exists z'(z' \preccurlyeq z \wedge A(z')) \wedge \Box\neg B)$ for every $\alpha \in \bar{\mathcal{M}}_\Pi$	(1), (2)
(6)	$A(z) \wedge \Box\neg B \to \Box\exists z'(z' \preccurlyeq z \wedge A(z'))$	(trans), (ind″), (4), (5)
(7)	$A(z) \wedge \Box\neg B \wedge \Box\neg\exists z'(z' \prec z \wedge A(z')) \to \Box A(z) \wedge \Box\neg B$	(pred), (6)
(8)	$\Box A(z) \to \Box\Diamond B \vee \Box\Diamond E^{(h(z))}$	(T41), (T47), (3)
(9)	$\Box\Diamond E^{(h(z))} \to \Diamond(\beta_1 \vee \ldots \vee \beta_m)$	(B3′)
(10)	$A(z) \wedge \Box\neg B \wedge \Box\neg\exists z'(z' \prec z \wedge A(z')) \to \Diamond((\beta_1 \vee \ldots \vee \beta_m) \wedge A(z) \wedge \Box\neg B)$	(T39), (7), (8), (9)
(11)	$\beta_i \wedge A(z) \wedge \Box\neg B \to \circ\exists z'(z' \prec z \wedge A(z'))$ for every $i = 1, \ldots, m$	(2)
(12)	$A(z) \wedge \Box\neg B \wedge \Box\neg\exists z'(z' \prec z \wedge A(z')) \to \Diamond\exists z'(z' \prec z \wedge A(z'))$	(chain), (som), (10), (11)
(13)	$A(z) \to \Diamond(B \vee \exists z'(z' \prec z \wedge A(z')))$	(12)
(14)	$\exists z A(z) \to \Diamond B$	(wfo), (13) □

Again, we note that this proof rule is derived by purely logical means from the basic program axioms (B2) and (B3). We only used (B3′) instead of (B3). The simple derivation of (B3′) from (B3) was shown in Section 20. Summarizing this together with the analogous observation in the previous section, we may state that:

- the general methods for proving eventuality properties expressed in (chain), (event0) and (well) are based on purely temporal logical laws and the two basic axioms (B2) and (B3).

Together with the statement at the end of Section 18 we have that:

- all our general proof methods for program properties are based on purely temporal logical laws and the basic program axioms (B1), (B2) and (B3).

We conclude this section with a simple illustration of how this general method of well-founded orderings may be applied.

Example. The program:

$$
\begin{array}{l}
\Pi \equiv \mathbf{initial}\ a \geq 0 \wedge b=1; \\
\quad \mathbf{cobegin}\ \alpha_0: \mathbf{while}\ a>0\ \mathbf{do}\ \alpha_1: a:=a-1\ \mathbf{od}; \\
\quad\quad\quad\quad \alpha_2: b:=0; \\
\quad\quad\quad\quad \alpha_3: \mathbf{stop} \\
\quad\quad\quad\quad \| \\
\quad\quad\quad\quad \beta_0: \mathbf{while}\ b=1\ \mathbf{do}\ \beta_1: c:=0\ \mathbf{od}; \\
\quad\quad\quad\quad \beta_2: \mathbf{stop} \\
\quad \mathbf{coend}
\end{array}
$$

certainly terminates. In particular we claim that:

$$\mathrm{start}_\Pi \rightarrow \Diamond\ \mathrm{at}\ \beta_2.$$

(Observe the similarity with the example in the previous section.) In order to prove the claim we let $Z=\mathbb{N}_0 \times \mathbb{N}_0$ with the lexicographical ordering, $z=(z_1, z_2)$, and $h(z)=1$ for every $z \in Z$. We consider the while loop of the first component. α_1 decreases the value of a which is represented by z_1 and z_2 is chosen in such a way that it is decreased by the entry to the loop body. More formally, we put:

$$A(z) \equiv [(\mathrm{at}\ \alpha_0 \wedge z_2=1) \vee (\mathrm{at}\ \alpha_1 \wedge a>0 \wedge z_2=0)] \wedge a=z_1$$

and get:

(1)	$\beta \wedge A(z) \rightarrow \circ(\mathrm{at}\ \alpha_2 \vee \exists z'(z' \preccurlyeq z \wedge A(z')))$ for $\beta \in \{\beta_0, \beta_1\}$	(Π)
(2)	$\alpha_0 \wedge A(z_1, z_2) \rightarrow \circ(\mathrm{at}\ \alpha_2 \vee A(z_1, z_2-1))$	(Π)
(3)	$\alpha_1 \wedge A(z_1, z_2) \rightarrow \circ(A(z_1-1, z_2+1))$	(Π)
(4)	$\alpha_2 \wedge A(z) \rightarrow \circ \exists z'(z' \prec z \wedge A(z'))$	(taut)
(5)	$\alpha \wedge A(z) \rightarrow \circ(\mathrm{at}\ \alpha_2 \vee \exists z'(z' \prec z \wedge A(z')))$ for $\alpha \in \{\alpha_0, \alpha_1, \alpha_2\}$	(2), (3), (4)
(6)	$A(z) \rightarrow \Diamond(\mathrm{at}\ \alpha_0 \vee \mathrm{at}\ \alpha_1 \vee \mathrm{at}\ \alpha_2)$	(taut)
(7)	$\Box A(z) \rightarrow \Diamond(\mathrm{at}\ \alpha_0 \vee \mathrm{at}\ \alpha_1 \vee \mathrm{at}\ \alpha_2)$	(6)
(8)	$\exists z A(z) \rightarrow \Diamond\ \mathrm{at}\ \alpha_2$	(well), (1), (5), (7)
(9)	$\mathrm{start}_\Pi \rightarrow \exists z A(z)$	(pred)
(10)	$\mathrm{start}_\Pi \rightarrow \Diamond\ \mathrm{at}\ \alpha_2$	(8), (9)

From (10) the claim is proved in just the same way as in the example of the previous section. □

23. Examples of Applications

Again we want to illustrate our proof methods by some more realistic programs. Let us first recall the producer-consumer program ψ_3 from Example 17.3.

Program Example 23.1

```
ψ3 ≡ initial ex=true ∧ bf=0 ∧ be=n ∧ n>0;
     cobegin loop α0 : ~ } produce section
                       :
                       :
                  α1 : await be>0 then be:=be−1;
                  α2 : await ex=true then ex:=false;
                   γ : ~ } store section
                       :
                       :
                  α3 : ex:=true;
                  α4 : bf:=bf+1
             end
             ||
             loop β0 : await bf>0 then bf:=bf−1;
                  β1 : await ex=true then ex:=false;
                       :  } get section
                       :
                  β2 : ex:=true;
                  β3 : be:=be+1;
                       :  } consume section
                       :
             end
     coend
```

We now want to show that the producer eventually stores every produced object:

Assertion

$$\psi_3 \vdash \text{at } \alpha_1 \rightarrow \Diamond \text{ at } \gamma.$$

This assertion is somewhat stronger than the deadlock-freedom already shown in Example 17.3. In this case it follows relatively simply from this property. However, we have to be a little bit careful. We have to presuppose that the consumer does not halt in the get and consume sections, otherwise, of course, the producer could get blocked at α_1 trivially. So we let:

L = set of labels in get section,
M = set of labels in consume section,

and make the following presuppositions:

(1) at $L \rightarrow \Diamond$ at β_2
(2) at $M \rightarrow \Diamond$ at β_0.

Proof of the assertion. We first recall the four assertions already proved under Example 17.3:

(3) $\Box(\text{at } \alpha_1 \wedge \text{at } \beta_0 \rightarrow be>0 \vee bf>0)$

(4) $\Box(\text{at } \alpha_1 \wedge \text{at } \beta_1 \rightarrow be>0 \vee ex=\mathbf{true})$

(5) $\Box(\text{at } \alpha_2 \wedge \text{at } \beta_0 \rightarrow ex=\mathbf{true} \vee bf>0)$

(6) $\Box(\text{at } \alpha_2 \wedge \text{at } \beta_1 \rightarrow ex=\mathbf{true})$.

We also proved there:

(7) $\Box(be\geq 0 \wedge bf\geq 0)$.

Next we have:

(8) $\Box \text{ at } \alpha_1 \wedge \text{at } \beta \rightarrow \Diamond(be>0)$ for every $\beta \in \bar{\mathcal{M}}_{\text{cons}}$

(we let the indices cons and prod denote the consumer and producer component, respectively).

Derivation of (8) *for* $\beta \equiv \beta_3$

(8.1)	$\beta_3 \rightarrow \circ(be>0)$	(Π)
(8.2)	$\text{at } \beta_3 \rightarrow \Diamond(be>0)$	(event2), (8.1)
(8.3)	$\Box \text{ at } \alpha_1 \wedge \text{at } \beta_3 \rightarrow \Diamond(be>0)$	(8.2)

Derivation of (8) *for* $\beta \equiv \beta_2$

(8.4)	$\text{at } \beta_2 \rightarrow \Diamond \text{ at } \beta_3$	as (8.2)
(8.5)	$\Box \text{ at } \alpha_1 \wedge \text{at } \beta_2 \rightarrow \Diamond(be>0)$	(chain), (8.3), (8.4)

Derivation of (8) *for* $\beta \in L$

(8.6)	$\text{at } L \rightarrow \Diamond \text{ at } \beta_2$	(1)
(8.7)	$\Box \text{ at } \alpha_1 \wedge \text{at } L \rightarrow \Diamond(be>0)$	(chain), (8.5), (8.6)

Derivation of (8) *for* $\beta \equiv \beta_1$

(8.8)	$\alpha \wedge \text{at } \beta_1 \wedge \Box \text{ at } \alpha_1 \wedge be\leq 0 \rightarrow \circ(\Box \text{ at } \alpha_1 \wedge be\leq 0)$ for every $\alpha \in \bar{\mathcal{M}}_{\text{prod}}$	(Π)
(8.9)	$\beta_1 \wedge \Box \text{ at } \alpha_1 \wedge be\leq 0 \rightarrow \circ \text{ at } L$	(Π)
(8.10)	$\Box(\text{at } \beta_1 \wedge \Box \text{ at } \alpha_1 \wedge be\leq 0) \rightarrow ex=\mathbf{true}$	(4)
(8.11)	$\text{at } \beta_1 \wedge \Box \text{ at } \alpha_1 \wedge be\leq 0 \rightarrow \Diamond \text{ at } L$	(event1), (8.8)–(8.10)
(8.12)	$\Box \text{ at } \alpha_1 \wedge \text{at } \beta_1 \wedge be\leq 0 \rightarrow \Diamond(be>0)$	(chain), (8.7), (8.11)
(8.13)	$\Box \text{ at } \alpha_1 \wedge \text{at } \beta_1 \wedge be>0 \rightarrow \Diamond(be>0)$	trivial
(8.14)	$\Box \text{ at } \alpha_1 \wedge \text{at } \beta_1 \wedge \Diamond(be>0)$	(8.12), (8.13)

Derivation of (8) *for* $\beta \equiv \beta_0$

(8.15)	$\alpha \wedge \text{at } \beta_0 \wedge \Box \text{ at } \alpha_1 \wedge bf>0 \rightarrow \circ(\Box \text{ at } \alpha_1 \wedge bf>0)$ for every $\alpha \in \bar{\mathcal{M}}_{\text{prod}}$	(Π)
(8.16)	$\beta_0 \wedge \Box \text{ at } \alpha_1 \wedge bf>0 \rightarrow \circ \text{ at } \beta_1$	(Π)

(8.17)	$\Box(\text{at } \beta_1 \wedge \Box \text{ at } \alpha_1 \wedge bf>0) \rightarrow \Box(bf>0)$	trivial
(8.18)	$\text{at } \beta_0 \wedge \Box \text{ at } \alpha_1 \wedge bf>0 \rightarrow \Diamond \text{ at } \beta_1$	(event1), (8.15)–(8.17)
(8.19)	$\text{at } \beta_0 \wedge \Box \text{ at } \alpha_1 \wedge bf>0 \rightarrow \Diamond (be<0)$	(chain), (8.14), (8.18)
(8.20)	$\Box \text{ at } \alpha_1 \wedge \text{at } \beta_0 \rightarrow \Diamond (be>0)$	(8.19), (3)

Derivation of (8) *for* $\beta \in M$

(8.21)	$\text{at } M \rightarrow \Diamond \text{ at } \beta_0$	(2)
(8.22)	$\Box \text{ at } \alpha_1 \wedge \text{at } M \rightarrow \Diamond (be>0)$	(chain), (8.20), (8.21)

Now we have the trivial invariant:

(9) $\Box(\text{at } \beta_0 \vee \text{at } \beta_1 \vee \text{at } \beta_2 \vee \text{at } \beta_3 \vee \text{at } L \vee \text{at } M)$

and get:

(10) $\Box \text{ at } \alpha_1 \rightarrow \Diamond (be>0)$

from (8) and (9). From (10) we reach our first goal:

(11) $\text{at } \alpha_1 \rightarrow \Diamond \text{ at } \alpha_2.$

Derivation of (11)

(11.1)	$\alpha_1 \rightarrow \bigcirc \text{ at } \alpha_2$	(Π)
(11.2)	$\text{at } \alpha_1 \rightarrow \Diamond \text{ at } \alpha_2$	(event2), (10), (11.1)

In just the same way one can prove (complete derivations are left as an exercise):

(12) $\Box \text{ at } \alpha_2 \wedge \text{at } \beta \rightarrow \Diamond (ex = \texttt{true})$ for every $\beta \in \bar{\mathcal{M}}_{\text{cons}}$

(13) $\Box \text{ at } \alpha_2 \rightarrow \Diamond (ex = \texttt{true})$

(14) $\text{at } \alpha_2 \rightarrow \Diamond \text{ at } \gamma.$

The assertion then follows from (11) and (14) with (chain). □

Our next example deals with termination of a non-cyclic program.

Program Example 23.2

$\psi_7 \equiv$ **initial** $a_1 = n \wedge a_2 = 0 \wedge a_3 = 1 \wedge 0 \le k \le n;$

cobegin α_0 : **while** $a_1 \neq n-k$ **do**
$\alpha_1 : a_3 := a_3 \times a_1;$
$\alpha_2 : a_1 := a_1 - 1$ **od**;
α_3 : **stop**
‖
β_0 : **while** $a_2 \neq k$ **do**
$\beta_1 : a_2 := a_2 + 1;$
β_2 : **await** $a_1 + a_2 \le n;$
$\beta_3 : a_3 := a_3 / a_2$ **od**;
β_4 : **stop**
coend

This program computes the binomial coefficient $\binom{n}{k} = \frac{n \times (n-1) \times \ldots \times (n-k+1)}{1 \times 2 \times \ldots \times k}$ for $n, k \in \mathbb{N}_0$. The first parallel component accumulates the multiplications with n, $n-1$, etc. and the second component accumulates the divisions by 1, 2, etc. The await statement is included in order to guarantee that the division of the current values of a_3 and a_2 yields an integer value. (We do not care about how this is mathematically assured by $a_1 + a_2 \leq n$.) We only want to show termination of ψ_7.

Assertion

$$\psi_7 \vdash \text{start}_{\psi_7} \rightarrow \Diamond(\text{at } \alpha_3 \wedge \text{at } \beta_4).$$

Proof of the assertion. As usual we begin with listing some global invariants:

(1) $\Box(\text{at } \beta_1 \rightarrow a_2 < k)$
(2) $\Box(\text{at } \beta_2 \rightarrow a_2 \leq k)$
(3) $\Box(\text{at } \alpha_3 \rightarrow a_1 = n-k)$
(4) $\Box(\text{at } \alpha_2 \rightarrow a_1 > n-k)$.

(1) and (2) are proved by (inv′) with the invariant:

$$(\text{at } \beta_1 \rightarrow a_2 < k) \wedge (\neg \text{ at } \beta_1 \rightarrow a_2 \leq k).$$

(3) and (4) run in the same way with the invariant:

$$(\text{at } \alpha_3 \rightarrow a_1 = n-k) \wedge (\text{at } \alpha_1 \vee \text{at } \alpha_2 \rightarrow a_1 > n-k).$$

Now let $Z = \mathbb{N}_0 \times \mathbb{N}_0 \times \mathbb{N}_0$ with lexicographical ordering, $z = (z_1, z_2, z_3)$,

$$h(z) = \begin{cases} 1, & \text{if } z_2 \neq 0, \\ 2 & \text{else,} \end{cases}$$

and

$$\begin{aligned} A(z) \equiv\ & z_1 = a_1 + k - a_2 \wedge \\ & (\text{at } \alpha_0 \leftrightarrow z_2 = 3) \wedge (\text{at } \alpha_1 \leftrightarrow z_2 = 2) \wedge (\text{at } \alpha_2 \leftrightarrow z_2 = 1) \wedge \\ & (\text{at } \alpha_3 \leftrightarrow z_2 = 0) \wedge (\text{at } \beta_0 \leftrightarrow z_3 = 2) \wedge (\text{at } \beta_1 \leftrightarrow z_3 = 1) \wedge \\ & (\text{at } \beta_2 \leftrightarrow z_3 = 4) \wedge (\text{at } \beta_3 \leftrightarrow z_3 = 3) \wedge (\text{at } \beta_4 \leftrightarrow z_3 = 0). \end{aligned}$$

Of course, we have:

(5) $\text{start}_{\psi_7} \rightarrow \exists z A(z)$

(taking $z = (n+k, 3, 2)$) and furthermore we prove:

(6) $\exists z A(z) \rightarrow \Diamond(\text{at } \alpha_3 \wedge \text{at } \beta_4)$

which, together with (5), proves the assertion.

Derivation of (6). Let Π_1 and Π_2 denote the two parallel components of ψ_7.

(6.1) $\beta_0 \wedge A(z_1, z_2, z_3) \wedge z_2 \neq 0 \rightarrow \circ A(z_1, z_2, z_3')$
with $z_3' \in \mathbb{N}_0$, $z_3' < z_3$ $\qquad (\Pi)$

(6.2) $\beta_1 \wedge A(z_1, z_2, z_3) \wedge z_2 \neq 0 \rightarrow \circ A(z_1 - 1, z_2, z_3')$
with $z_1 - 1 \in \mathbb{N}_0$ $\qquad (\Pi)$, (1)

(6.3)	$\beta_2 \wedge A(z_1, z_2, z_3) \wedge z_2 \neq 0 \rightarrow \circ A(z_1, z_2, z_3 - 1)$ with $z_3 - 1 \in \mathbb{N}_0$	(Π)
(6.4)	$\beta_3 \wedge A(z_1, z_2, z_3) \wedge z_2 \neq 0 \rightarrow \circ A(z_1, z_2, z_3 - 1)$ with $z_3 - 1 \in \mathbb{N}_0$	(Π)
(6.5)	$\alpha_0 \wedge A(z_1, z_2, z_3) \wedge z_2 \neq 0 \rightarrow \circ A(z_1, z_2', z_3)$ with $z_2' \in \mathbb{N}_0, z_2' < z_2$	(Π)
(6.6)	$\alpha_1 \wedge A(z_1, z_2, z_3) \wedge z_2 \neq 0 \rightarrow \circ A(z_1, z_2 - 1, z_3)$ with $z_2 - 1 \in \mathbb{N}_0$	(Π)
(6.7)	$\alpha_2 \wedge A(z_1, z_2, z_3) \wedge z_2 \neq 0 \rightarrow \circ A(z_1 - 1, z_2', z_3)$ with $z_1 - 1 \in \mathbb{N}_0$	(Π), (3)
(6.8)	$A(z) \wedge z_2 \neq 0 \rightarrow \text{at } \alpha_0 \vee \text{at } \alpha_1 \vee \text{at } \alpha_2$	(pred)
(6.9)	$A(z) \wedge z_2 \neq 0 \rightarrow \Diamond [(\text{at } \alpha_3 \wedge \text{at } \beta_4) \vee E^{(h(z))}]$	(6.8)
(6.10)	$A(z) \wedge z_2 = 0 \rightarrow \text{at } \alpha_3$	(pred)
(6.11)	$\beta \wedge A(z) \wedge z_2 = 0 \rightarrow \circ [(\text{at } \alpha_3 \wedge \text{at } \beta_4) \vee \exists z'(z' \prec z \wedge A(z'))]$ for every $\beta \in \bar{\mathcal{M}}_{\Pi_2}$	(as above)
(6.12)	$A(z) \wedge z_2 = 0 \wedge (\text{at } \beta_0 \vee \text{at } \beta_1 \vee \text{at } \beta_3 \vee \text{at } \beta_4) \rightarrow (\text{at } \alpha_3 \wedge \text{at } \beta_4) \vee (\text{at } \beta_0 \vee \text{at } \beta_1 \vee \text{at } \beta_3)$	(6.10)
(6.13)	$A(z) \wedge z_2 = 0 \wedge \text{at } \beta_2 \rightarrow \text{at } \alpha_3 \wedge \text{at } \beta_2$	(6.10)
(6.14)	$\text{at } \alpha_3 \wedge \text{at } \beta_2 \rightarrow a_1 + a_2 \leq n$	(2), (3)
(6.15)	$A(z) \wedge z_2 = 0 \rightarrow \Diamond [(\text{at } \alpha_3 \wedge \text{at } \beta_4) \vee E^{(h(z))}]$	(6.12)–(6.14)
(6.16)	$\gamma \wedge A(z) \rightarrow \circ [(\text{at } \alpha_3 \wedge \text{at } \beta_4) \vee \exists z'(z' \preceq z \wedge A(z'))]$ for every $\gamma \in \bar{\mathcal{M}}_{\psi_7} \setminus \bar{\mathcal{M}}_{\Pi_{h(z)}}$	(6.1)–(6.4), (6.10)
(6.17)	$\gamma \wedge A(z) \rightarrow \circ [(\text{at } \alpha_3 \wedge \text{at } \beta_4) \vee \exists z'(z' \prec z \wedge A(z'))]$ for every $\gamma \in \bar{\mathcal{M}}_{\Pi_{h(z)}}$	(6.5)–(6.7), (6.11)
(6.18)	$\Box A(z) \rightarrow \Diamond [(\text{at } \alpha_3 \wedge \text{at } \beta_4) \vee E^{(h(z))}]$	(6.9), (6.15)
(6.19)	$\exists z A(z) \rightarrow \Diamond (\text{at } \alpha_3 \wedge \text{at } \beta_4)$	(well), (6.16)–(6.18) □

Our third application deals again with the alternating bit protocol ψ_4 discussed in Example 19.1.

Program Example 23.3

```
ψ4 ≡ initial nextinput = u0 ∧ nr = 1 ∧ ls = mn = a = 0;
     cobegin loop α0: if ls = a then ls := ls⊕1;
                                     d := nextinput fi;
                      α1: send(ls, d) to (mn, inf)
             end
             ‖
             loop β0: if mn = nr then nextoutput := inf;
                                      nr := nr⊕1 fi;
                      β1: send(nr⊕1) to (a)
             end
     coend
```

We already proved under Example 19.1 that if messages arrive at all then their arrival and realization in β_0 is correct. We now want to show that in fact infinitely many messages are realized in β_0. Altogether this means that all input messages are transferred correctly.

Assertion

$$\psi_4 \vdash \Diamond(\beta_0 \wedge mn = nr).$$

This assertion is not exactly what is expressed verbally above but it is easily seen to be equivalent to a more direct transcription $\psi_4 \vdash \mathrm{start}_{\psi_4} \rightarrow \Box\Diamond(\beta_0 \wedge mn = nr)$. However, as the program ψ_4 stands, the assertion is trivially false. We have not excluded the case that sending operations always fails and, in particular, mn will (after the first execution of α_1) always have the "value" error. The eventual arrival of messages (and acknowledgements) can only be guaranteed if (at every time) some eventual sending operation will provide uncorrupted values in the respective mail boxes mn and a. Moreover, these values must be read by β_0 and α_0, respectively, before a new corrupted transmission arrives. Formally we may note these conditions by the presuppositions:

(1) $\Diamond(\alpha_0 \wedge a \neq \mathrm{error})$
(2) $\Diamond(\beta_0 \wedge mn \neq \mathrm{error})$

which we assume to be guaranteed in addition to the program text of ψ_4.

Proof of the assertion. We first notice that "parts" of (1) and (2) could be derived from our general fairness assumption, namely that α_0 and β_0 are eventually executed at all (i.e., $\Diamond\alpha_0$ and $\Diamond\beta_0$). We also have to derive this for α_1 and β_1:

(3) $\Diamond\alpha_1$
(4) $\Diamond\beta_1$.

It is remarkable that the proofs of (3) and (4) need none of our rules discussed in Sections 21 and 22 but run best by directly applying the fairness axiom (B3):

Derivation of (3). Observe that $E_{\alpha_1} \equiv \mathtt{true}$.

(3.1)	$\Diamond\alpha_0$	(1)
(3.2)	$\alpha_0 \rightarrow \circ\ \mathrm{at}\ \alpha_1$	(Π)
(3.3)	$\Diamond\ \mathrm{at}\ \alpha_1$	(3.1), (3.2)
(3.4)	$\Box\Diamond\ \mathrm{at}\ \alpha_1$	(alw), (3.3)
(3.5)	$\Diamond\ \alpha_1$	(B3), (3.4)

The derivation of (4) is analogous.

Next we note two formulas proved already under Example 19.1:

(5) $\Box((nr=0 \vee nr=1) \wedge (ls=0 \vee ls=1))$
(6) $\Box((mn=nr \rightarrow nr=ls \wedge inf=d) \wedge (ls=a \rightarrow nr \neq ls))$

and then prove:

(7) $\Box(\beta_0 \rightarrow mn \neq nr) \rightarrow \Diamond(nr=ls)$.

Derivation of (7)

(7.1)	$\beta_1 \to \circ(a = nr \oplus 1 \vee a = \text{error})$	(Π)
(7.2)	$\Diamond(a = nr \oplus 1 \vee a = \text{error})$	(som), (4), (7.1)
(7.3)	$\Box(\beta_0 \to mn \neq nr) \wedge \Box(nr \neq ls) \to$ $\Diamond[(a = nr \oplus 1 \vee a = \text{error}) \wedge \Box(\beta_0 \to mn \neq nr) \wedge \Box(nr \neq ls)]$	(7.2)
(7.4)	$(a = nr \oplus 1 \vee a = \text{error}) \wedge \Box(\beta_0 \to mn \neq nr) \wedge \Box(nr \neq ls) \to$ $\Box(a = nr \oplus 1 \vee a = \text{error}) \wedge \Box(\beta_0 \to mn \neq nr) \wedge \Box(nr \neq ls)$	(inv)
(7.5)	$\Box(a = nr \oplus 1 \vee a = \text{error}) \wedge \Box(nr \neq ls) \to \Box(a = ls \vee a = \text{error})$	(5)
(7.6)	$\Box(a = ls \vee a = \text{error}) \to$ $\Diamond(\alpha_0 \wedge a \neq \text{error} \wedge (a = ls \vee a = \text{error}))$	(1)
(7.7)	$\Box(\beta_0 \to mn \neq nr) \wedge \Box(nr \neq ls) \to \Diamond(\alpha_0 \wedge a = ls \wedge nr \neq ls)$	(7.3)–(7.6)
(7.8)	$\alpha_0 \wedge a = ls \wedge nr \neq ls \to \circ(nr = ls)$	(Π), (5)
(7.9)	$\Box(\beta_0 \to mn \neq nr) \wedge \Box(nr \neq ls) \to \Diamond(nr = ls)$	(chain), (som), (7.7), (7.8)
(7.10)	$\Box(\beta_0 \to mn \neq nr) \to \Diamond(nr = ls)$	(7.9)

Next, it is easily seen by (inv) and (6) that:

(8) $\Box(\beta_0 \to mn \neq nr) \wedge nr = ls \to \Box(nr = ls \wedge a \neq ls)$

from which we derive:

(9) $nr = ls \to \Diamond(\beta_0 \wedge mn = nr)$.

Derivation of (9)

(9.1)	$\alpha_1 \to \circ(mn = ls \vee mn = \text{error})$	(Π)
(9.2)	$\Diamond(mn = ls \vee mn = \text{error})$	(som), (3), (9.1)
(9.3)	$\Box(\beta_0 \to mn \neq nr) \wedge nr = ls \to \Diamond[(mn = ls \vee mn = \text{error}) \wedge$ $\Box(\beta_0 \to mn \neq nr) \wedge \Box(nr = ls \wedge a \neq ls)]$	(8), (9.2)
(9.4)	$(mn = ls \vee mn = \text{error}) \wedge \Box(\beta_0 \to mn \neq nr) \wedge$ $\Box(nr = ls \wedge a \neq ls) \to \Box(mn = ls \vee mn = \text{error}) \wedge$ $\Box(\beta_0 \to mn \neq nr) \wedge \Box(nr = ls \wedge a \neq ls)$	(inv)
(9.5)	$\Box(mn = ls \vee mn = \text{error}) \wedge \Box(nr = ls) \to$ $\Box(mn = nr \vee mn = \text{error})$	(pred)
(9.6)	$\Box(mn = nr \vee mn = \text{error}) \to$ $\Diamond(\beta_0 \wedge mn \neq \text{error} \wedge (mn = nr \vee mn = \text{error}))$	(2)
(9.7)	$\Box(\beta_0 \to mn \neq nr) \wedge nr = ls \to \Diamond(\beta_0 \wedge mn = nr)$	(9.3)–(9.6)
(9.8)	$nr = ls \to \Diamond(\beta_0 \wedge mn = nr)$	(9.7)

By (chain) we now find:

(10) $\Box(\beta_0 \to mn \neq nr) \to \Diamond(\beta_0 \wedge mn = nr)$

from (7) and (9) and this yields the assertion immediately with (prop). □

Chapter VII
Special Methods for Sequential Programs

24. Hoare's Calculus

All proof principles discussed in the previous sections are applicable to any program. For sequential programs there exist two further special verification methods which – although related to the universal methods – have some significance of their own and are sometimes easier to use. In this last chapter we want to represent these two methods within our temporal logic framework.

Thus, for the rest of this book, let Π be a non-cyclic sequential program:

$$\begin{array}{lll} \Pi \equiv \texttt{initial}\ R; & & \\ \quad \alpha_0: & \vdots & \\ & \vdots & \\ \quad \alpha_e: & \texttt{stop} & \end{array}$$

containing, of course, no synchronization statements. (We also assume $\alpha_0 \not\equiv \alpha_e$.)

The first method we want to describe is a method for proving the partial correctness (w.r.t. some P and Q) for such a program. This property is expressed by the formula:

$$\text{start}_\Pi \wedge P \rightarrow \Box(\text{at } \alpha_e \rightarrow Q).$$

The general method to prove such a formula is the basic invariant method directly induced by the induction principle (ind″). We now recall another (generalized) induction principle derived in Section 7:

(gind) $\quad A \rightarrow C\ \texttt{atnext}\ B,\ B \wedge C \rightarrow C\ \texttt{atnext}\ B \vdash A \rightarrow \Box\circ(B \rightarrow C).$

(gind) gives us another possibility for proving the partial correctness formula above, expressed by the following rule:

(H0) $\quad \text{start}_\Pi \wedge P \rightarrow Q\ \texttt{atnext}\ \text{at } \alpha_e \vdash \text{start}_\Pi \wedge P \rightarrow \Box(\text{at } \alpha_e \rightarrow Q).$

This rule should be intuitively clear since if Q holds upon the first arrival at α_e nothing is changed anymore. We also give a formal derivation.

Derivation of (H0)

(1)	$\text{start}_\Pi \wedge P \rightarrow Q\ \texttt{atnext}\ \text{at } \alpha_e$	assumption
(2)	$\text{at } \alpha_e \wedge Q \rightarrow \text{nil}_\Pi$	(Π)
(3)	$\text{at } \alpha_e \wedge Q \rightarrow \circ(\text{at } \alpha_e \wedge Q)$	(Π), (2)
(4)	$\circ(\text{at } \alpha_e \wedge Q) \rightarrow Q\ \texttt{atnext}\ \text{at } \alpha_e$	(prop), (ax 5)
(5)	$\text{at } \alpha_e \wedge Q \rightarrow Q\ \texttt{atnext}\ \text{at } \alpha_e$	(3), (4)

(6)	$\text{start}_\Pi \wedge P \rightarrow \Box\bigcirc(\text{at } \alpha_e \rightarrow Q)$	(gind), (1), (5)
(7)	$\text{start}_\Pi \rightarrow \neg \text{ at } \alpha_e$	(Π)
(8)	$\text{start}_\Pi \wedge P \rightarrow \Box(\text{at } \alpha_e \rightarrow Q)$	(6), (7) □

Consider now the general situation of a program "piece" ψ (formally: a statement subsequence) of Π with entry$(\psi)=\beta_1$ (for the notion of entry(ψ), cf. Section 13):

$$\beta_1 : \left.\begin{matrix}\vdots\\ \vdots\end{matrix}\right\}\psi$$

We define the label after(ψ) inductively by considering the program part ψ' which "immediately contains" ψ:

1. If ψ is the entire statement sequence of Π (without the **stop** statement) then after $(\psi)=\alpha_e$.
2. $\psi' \equiv \psi; \psi''$: after$(\psi)$=entry$(\psi'')$.
3. $\psi' \equiv \psi''; \psi$: after$(\psi)$=after$(\psi')$.
4. $\psi' \equiv$ **if** B **then** ψ **else** ψ'' **fi**: after(ψ)=after(ψ').
5. $\psi' \equiv$ **if** B **then** ψ'' **else** ψ **fi**: after(ψ)=after(ψ').
6. $\psi' \equiv$ **if** B **then** ψ **fi**: after(ψ)=after(ψ').
7. $\psi' \equiv$ **while** B **do** ψ **od**: after(ψ)=entry(ψ').

Informally, after(ψ) is that label where computation continues after executing ψ.

Example. Let

```
Π ≡ initial R;
    α0: while B do α1: ~;
                   α2: ~od;
    α3: stop
```

Then we compute:

after(α_0: **while** B **do** ... **od**)$=\alpha_3$,
after$(\alpha_1 : \sim)=\alpha_2$,
after$(\alpha_2 : \sim)$=after$(\alpha_1 : \sim; \alpha_2 : \sim)=\alpha_0$. □

We are now interested in formulas of the kind:

$$\text{at } \beta_1^\psi \wedge P \rightarrow Q \textbf{ atnext } \text{at } \beta_2^\psi$$

where β_1^ψ=entry(ψ), β_2^ψ=after(ψ), and P and Q are P-formulas, and use a new notation for them:

$$P[\beta_1^\psi, \beta_2^\psi]\, Q.$$

The fact that $\beta_1^\psi=\alpha_0$ and $\beta_2^\psi=\alpha_e$ if ψ is the entire statement sequence of Π allows us to summarize this discussion to:

Theorem 24.1. *If* $\Pi \vdash P \wedge R[\alpha_0, \alpha_e]\, Q$ *then* Π *is partially correct w.r.t.* P *and* Q *(i.e.,* $\Pi \vdash \text{start}_\Pi \wedge P \rightarrow \Box(\text{at } \alpha_e \rightarrow Q)$*).*

Proof. $P \wedge R[\alpha_0, \alpha_e]\, Q$ means at $\alpha_0 \wedge R \wedge P \rightarrow Q$ **atnext** at α_e and applying rule (H0) yields the assertion. □

The crucial point now is that the formula $P \wedge R[\alpha_0, \alpha_e]\, Q$ need not be proved by the general proof method with, say, the rule (atnext). There is a particular set of simple rules for – generally – proving formulas of the form $P[\beta_1^\psi, \beta_2^\psi]\, Q$ (*Hoare's calculus*). Each of these rules (except two "technical" ones) deals with one syntactical structure of ψ and states in the case of a compound ψ how to derive $P[\beta_1^\psi, \beta_2^\psi]\, Q$ from the corresponding formulas for the constituent(s) of ψ. (For simplicity we write β_1 and β_2 instead of β_1^ψ and β_2^ψ.)

(H1) $\beta_1 \wedge P \rightarrow \circ(\text{at } \beta_2 \wedge Q) \vdash P[\beta_1, \beta_2]\, Q$ if ψ is an elementary statement

(H2) $P[\beta_1, \gamma]\, P', P'[\gamma, \beta_2]\, Q \vdash P[\beta_1, \beta_2]\, Q$
if $\psi \equiv \psi_1 ; \psi_2$, $\gamma = \text{after}(\psi_1) = \text{entry}(\psi_2)$

(H3) $P \wedge B[\gamma_1, \beta_2]\, Q, P \wedge \neg B[\gamma_2, \beta_2]\, Q \vdash P[\beta_1, \beta_2]\, Q$
if $\psi \equiv$ **if** B **then** ψ_1 **else** ψ_2 **fi**, $\gamma_1 = \text{entry}(\psi_1)$, $\gamma_2 = \text{entry}(\psi_2)$

(H4) $P \wedge B[\gamma_1, \beta_2]\, Q,\ P \wedge \neg B \rightarrow Q \vdash P[\beta_1, \beta_2]\, Q$
if $\psi \equiv$ **if** B **then** ψ_1 **fi**, $\gamma_1 = \text{entry}(\psi_1)$

(H5) $P \wedge B[\gamma_1, \beta_1]\, P \vdash P[\beta_1, \beta_2]\, P \wedge \neg B$
if $\psi \equiv$ **while** B **do** ψ_1 **od**, $\gamma_1 = \text{entry}(\psi_1)$

(H6) $P[\beta_1, \beta_2]\, P',\ P' \rightarrow Q \vdash P[\beta_1, \beta_2]\, Q$

(H7) $P \rightarrow P',\ P'[\beta_1, \beta_2]\, Q \vdash P[\beta_1, \beta_2]\, Q$

The most interesting rule is the *loop rule* (H5). The formula P in (H5) is called *loop invariant*. Its informal meaning is:

> "If P holds at β_1 then P holds again when returning at β_1 after one loop run".

(H5) states that if P is such a loop invariant and P holds at the beginning of the loop then $P \wedge \neg B$ holds whenever the loop terminates.

Observe that only (H1) actually reminds us of temporal logic. In the standard case, when the elementary statements are assignments, we could replace (H1) by the axiom (scheme):

(H 1′) $P_a(t)[\beta_1, \beta_2]\,P$ if ψ is an assignment $a := t$

which follows immediately with (assign). Then the language of formulas $P[\beta_1^\psi, \beta_2^\psi]\,Q$ can be viewed as a separate language and one need not know of its temporal logic interpretation. We could even drop the assumption that statements are labelled since β_1^ψ and β_2^ψ are uniquely determined by ψ and we also could write $P[\psi]\,Q$ instead of $P[\beta_1^\psi, \beta_2^\psi]\,Q$. This is, in fact, the usual (non-temporal) way of presenting this partial correctness logic. In our context, where we have presupposed a program syntax with labelled statements, we prefer the notation with labels since it is much shorter. The embedding of the calculus in our temporal logic framework enables us to give immediate justifications of the rules.

Theorem 24.2. *The rules (H 1)–(H 7) are Π-derived.*

Proof. We give a direct derivation of (H 1):

(1)	$\beta_1 \wedge P \rightarrow \circ(\text{at } \beta_2 \wedge Q)$	assumption
(2)	$\text{at } \beta_1 \wedge P \rightarrow \beta_1 \wedge P$	(Π)
(3)	$\text{at } \beta_1 \wedge P \rightarrow \circ(\text{at } \beta_2 \rightarrow Q) \wedge \circ(\neg \text{ at } \beta_2 \rightarrow \text{at } \beta_1 \wedge P)$	(1), (2)
(4)	$\text{at } \beta_1 \wedge P \rightarrow Q \;\mathbf{atnext}\; \text{at } \beta_2$	(indatnext), (3)

(H 2)–(H 5) are also proved with (indatnext). The derivations are very similar to each other. We show the complete derivation of (H 2). Let $A \equiv A_1 \vee A_2$ with $A_1 \equiv \text{at } \mathcal{M}_{\psi_1} \wedge P' \;\mathbf{atnext}\; \text{at } \gamma$ and $A_2 \equiv \text{at } \mathcal{M}_{\psi_2} \wedge Q \;\mathbf{atnext}\; \text{at } \beta_2$. Then we have:

(1)	$\text{at } \beta_1 \wedge P \rightarrow P' \;\mathbf{atnext}\; \text{at } \gamma$	assumption
(2)	$\text{at } \gamma \wedge P' \rightarrow Q \;\mathbf{atnext}\; \text{at } \beta_2$	assumption
(3)	$A_1 \rightarrow \circ(\text{at } \gamma \wedge P') \vee \circ(\neg \text{ at } \gamma \wedge \text{at } \mathcal{M}_{\psi_1} \wedge P' \;\mathbf{atnext}\; \text{at } \gamma)$	(Π), (ax 5)
(4)	$A_1 \rightarrow \circ(\text{at } \gamma \wedge Q \;\mathbf{atnext}\; \text{at } \beta_2) \vee \circ A_1$	(2), (3)
(5)	$A_1 \rightarrow \circ(\neg \text{ at } \beta_2 \wedge A)$	(Π), (4)
(6)	$A_2 \rightarrow \circ(\text{at } \beta_2 \rightarrow Q) \wedge$	
	$\circ(\neg \text{ at } \beta_2 \rightarrow \text{at } \mathcal{M}_{\psi_2} \wedge Q \;\mathbf{atnext}\; \text{at } \beta_2)$	(Π), (ax 5)
(7)	$A_2 \rightarrow \circ(\text{at } \beta_2 \rightarrow Q) \wedge \circ(\neg \text{ at } \beta_2 \rightarrow A)$	(6)
(8)	$A \rightarrow \circ(\text{at } \beta_2 \rightarrow Q) \wedge \circ(\neg \text{ at } \beta_2 \rightarrow A)$	(5), (7)
(9)	$A \rightarrow Q \;\mathbf{atnext}\; \text{at } \beta_2$	(indatnext), (8)
(10)	$\text{at } \beta_1 \wedge P \rightarrow A$	(1)
(11)	$\text{at } \beta_1 \wedge P \rightarrow Q \;\mathbf{atnext}\; \text{at } \beta_2$	(9), (10)

For (H 3) one proves separately:

(1) $\text{at } \beta_1 \wedge P \wedge B \rightarrow Q \;\mathbf{atnext}\; \text{at } \beta_2$

by (indatnext) with $A \equiv (\text{at } \beta_1 \wedge P \wedge B) \vee (\text{at } \mathcal{M}_{\psi_1} \wedge Q \;\mathbf{atnext}\; \text{at } \beta_2)$ and:

(2) $\text{at } \beta_1 \wedge P \wedge \neg B \rightarrow Q \;\mathbf{atnext}\; \text{at } \beta_2$

again with (indatnext) and $A \equiv (\text{at } \beta_1 \wedge P \wedge \neg B) \vee (\text{at } \mathcal{M}_{\psi_2} \wedge Q \text{ } \mathbf{atnext} \text{ at } \beta_2)$. In (H 4) the first part is the same, the second is trivial. In both cases we get $\text{at } \beta_1 \wedge P \rightarrow Q$ **atnext** at β_2 from (1) and (2).
The proof of (H 5) runs analogously with:

$$A \equiv (\text{at } \beta_1 \wedge P) \vee (\text{at } \mathcal{M}_{\psi_1} \wedge P \text{ } \mathbf{atnext} \text{ at } \beta_1).$$

(H 6), finally, follows immediately with (T 36), and (H 7) is trivial. □

Let us now illustrate the use of Hoare's calculus by a simple example.

Example. Let

$$\begin{aligned} \Pi \equiv\ & \mathbf{initial}\ a=1 \wedge b=n \wedge n \geq 0; \\ & \alpha_0: \mathbf{while}\ b \neq 0\ \mathbf{do}\ \alpha_1: a:=2 \times a; \\ & \qquad\qquad\qquad\qquad \alpha_2: b:=b-1\ \mathbf{od}; \\ & \alpha_3: \mathbf{stop} \end{aligned}$$

The assertion is that Π is partially correct w.r.t. the (additional) precondition **true** and the postcondition $a=2^n$, i.e., that $\text{start}_\Pi \rightarrow \Box(\text{at } \alpha_3 \rightarrow a=2^n)$ holds. According to Theorem 24.1 we want to derive:

$$a=1 \wedge b=n \wedge n \geq 0 [\alpha_0, \alpha_3]\, a=2^n$$

by using (H 1)–(H 7). Let $P \equiv a=2^{n-b}$. We first show that this P is a loop invariant for the while loop in Π:

(1) $\quad P \wedge b \neq 0 [\alpha_1, \alpha_0]\, P.$

Derivation of (1)

(1.1)	$a=2^{n-b} \wedge b \neq 0 [\alpha_1, \alpha_2]\, a=2^{n-b+1}$	(Π), (H 1)
(1.2)	$a=2^{n-b+1} [\alpha_2, \alpha_0]\, a=2^{n-b}$	(Π), (H 1)
(1.3)	$a=2^{n-b} \wedge b \neq 0 [\alpha_1, \alpha_0]\, a=2^{n-b}$	(H 2), (1.1), (1.2)

From (1) we now get:

(2) $\quad P[\alpha_0, \alpha_3]\, P \wedge b=0$

with (H 5). Furthermore, we have:

(3) $\quad a=1 \wedge b=n \wedge n \geq 0 \rightarrow P$
(4) $\quad P \wedge b=0 \rightarrow a=2^n$

and so we get $a=1 \wedge b=n \wedge n \geq 0 [\alpha_0, \alpha_3]\, a=2^n$ with (H 6) and (H 7). □

The example shows one characteristic point of such a proof with (H 1)–(H 7). For any label of the program, we describe – with appropriate formulas – relationships between the variables. These relationships hold whenever the execution of the program is at that label. These formulas could also be put together to a global invariant; in our example:

$$\begin{aligned} A \equiv\ & (\text{at } \alpha_0 \rightarrow a=2^{n-b}) \wedge \\ & (\text{at } \alpha_1 \rightarrow a=2^{n-b} \wedge b \neq 0) \wedge \\ & (\text{at } \alpha_2 \rightarrow a=2^{n-b+1}) \wedge \\ & (\text{at } \alpha_3 \rightarrow a=2^n). \end{aligned}$$

This shows the strong relationship of Hoare's calculus with the basic invariant method discussed in Section 16. In fact, A could be used for a proof with (inv). We have $\text{start}_\Pi \rightarrow A$ and A **invof** $\{\alpha_0, \alpha_1, \alpha_2\}$ and hence may deduce $\text{start}_\Pi \rightarrow \Box(\text{at } \alpha_3 \rightarrow a = 2^n)$.

Thus, Hoare's calculus (which was actually proposed much earlier than temporal logic verification rules) does not really involve another proof "idea". Rather, it allows one to formulate the same idea which one would have for a proof with (inv) in a different way.

25. The Intermittent Assertion Method

We now turn to total correctness of programs. In general, one reasonable way of proving this property is to split it into separate proofs of partial correctness and termination according to the trivial rule:

$$\begin{array}{ll} \text{start}_\Pi \wedge P \rightarrow \Box(\text{at } \alpha_e^{(1)} \wedge \ldots \wedge \text{at } \alpha_e^{(p)} \rightarrow Q), & \text{(partial correctness)} \\ \text{start}_\Pi \wedge P \rightarrow \Diamond(\text{at } \alpha_e^{(1)} \wedge \ldots \wedge \text{at } \alpha_e^{(p)}) & \text{(termination)} \\ \vdash \text{start}_\Pi \wedge P \rightarrow \Diamond(\text{at } \alpha_e^{(1)} \wedge \ldots \wedge \text{at } \alpha_e^{(p)} \wedge Q) & \text{(total correctness)} \end{array}$$

In fact, in the case of parallel programs, this way is often the simplest. For sequential programs, however, there exists a method for proving total correctness directly which is very simply applicable in many cases. (Of course, it may also be used to prove only termination.) The method, called the *intermittent assertion method*, is related to the method of well-founded orderings. Consider once more the simple program example of Section 22:

$$\begin{array}{l} \Pi \equiv \textbf{initial } a \geq 0; \\ \quad \alpha_0 : \textbf{while } a > 0 \textbf{ do } \alpha_1 : a := a - 1 \textbf{ od}; \\ \quad \alpha_2 : \textbf{stop} \end{array}$$

with the assertion that:

$$\text{at } \alpha_0 \wedge a \geq 0 \rightarrow \Diamond \text{at } \alpha_2.$$

The idea for proving this was to take a formula $A(n) \equiv \text{at } \alpha_0 \wedge a = n$ (n ranging over $\mathbb{N}_0$) with the properties:

$$\begin{array}{ll} \text{at } \alpha_0 \wedge a \geq 0 \rightarrow \exists n A(n) & \text{and} \\ A(n) \rightarrow \Diamond(B \vee A(n-1)) & \text{(writing } B \text{ for at } \alpha_2). \end{array}$$

By induction the latter formula provides $A(n) \rightarrow \Diamond B$ and the rest is trivial.

We could have formulated this same proof idea in other words:

Prove $A(n) \rightarrow \Diamond B$ by induction on n, i.e.,

$$\begin{array}{lll} \text{i)} & A(0) \rightarrow a = 0, & \text{hence by } (\Pi) \\ & A(0) \rightarrow \circ B, & \text{and therefore} \\ & A(0) \rightarrow \Diamond B; & \\ \text{ii)} & A(n+1) \rightarrow \Diamond A(n) & \text{by } (\Pi), \\ & A(n) \rightarrow \Diamond B & \text{by induction hypothesis, and therefore} \\ & A(n+1) \rightarrow \Diamond B. & \end{array}$$

Of course, it is essential for this simple induction that B does not contain the variable n. If n were contained in B then the induction hypothesis in ii) would read:

$$A(n) \rightarrow \Diamond B(n)$$

and from this we could not deduce $A(n+1) \rightarrow \Diamond B(n+1)$ in the same simple way as above. However, we may try to permit this case of n occurring in B and overcome the difficulty by more skilful inductive arguments. Let us consider as an example the program:

$$\begin{aligned} \Pi \equiv\ & \mathbf{initial}\ a=1 \wedge b=n; \\ & \alpha_0 : \mathbf{while}\ b \neq 0\ \mathbf{do}\ \alpha_1 : a := 2 \times a; \\ & \qquad\qquad\qquad\qquad \alpha_2 : b := b-1\ \mathbf{od}; \\ & \alpha_3 : \mathbf{stop} \end{aligned}$$

already discussed in the previous section, this time with the total correctness assertion:

$$\text{start}_\Pi \rightarrow \Diamond(\text{at } \alpha_3 \wedge a=2^n).$$

In order to prove this assertion we prove another formula:

(1) $\quad \text{at } \alpha_0 \wedge a=a_0 \wedge b=n \rightarrow \Diamond(\text{at } \alpha_3 \wedge a=a_0 \times 2^n)$

where a_0 is a new variable. n is already a variable over $\mathbb{N}_0$ and we are able to prove (1) by induction on n.

i) $n=0$:

(1.1)	$\text{at } \alpha_0 \wedge a=a_0 \wedge b=n \wedge n=0 \rightarrow \circ(\text{at } \alpha_3 \wedge a=a_0 \wedge b=n \wedge n=0)$	(Π)
(1.2)	$a=a_0 \wedge b=n \wedge n=0 \rightarrow a=a_0 \times 2^n$	(data)
(1.3)	$\text{at } \alpha_0 \wedge a=a_0 \wedge b=0 \rightarrow \Diamond(\text{at } \alpha_3 \wedge a=a_0 \times 2^n)$	(1.1), (1.2)

ii) $n>0$:

(1.4)	$\text{at } \alpha_0 \wedge a=a_0 \wedge b=n \wedge n>0 \rightarrow \circ(\text{at } \alpha_1 \wedge a=a_0 \wedge b=n \wedge n>0)$	(Π)
(1.5)	$\text{at } \alpha_1 \wedge a=a_0 \wedge b=n \wedge n>0 \rightarrow \Diamond(\text{at } \alpha_0 \wedge a=2 \times a_0 \wedge b=n-1)$	(Π)
(1.6)	$\text{at } \alpha_0 \wedge a=2 \times a_0 \wedge b=n-1 \rightarrow \Diamond(\text{at } \alpha_3 \wedge a=(2 \times a_0) \times 2^{n-1})$	(ind.hyp.)
(1.7)	$a=(2 \times a_0) \times 2^{n-1} \rightarrow a=a_0 \times 2^n$	(data)
(1.8)	$\text{at } \alpha_0 \wedge a=a_0 \wedge b=n \wedge n>0 \rightarrow \Diamond(\text{at } \alpha_3 \wedge a=a_0 \times 2^n)$	(1)–(4)

Now we get:

(2) $\quad \text{at } \alpha_0 \wedge a=1 \wedge b=n \rightarrow \Diamond(\text{at } \alpha_3 \wedge a=2^n)$

from (1) by setting $a_0=1$ and (2) is just what we wanted to prove.

The general scheme of this proof is as follows:

1) If $A' \rightarrow \Diamond B'$ is to be proved then an appropriate "more general" formula $A \rightarrow \Diamond B$ is taken which possibly contains additional variables and from which $A' \rightarrow \Diamond B'$ results by "specializing" these variables. (A and B are called *intermittent assertions.*)
2) The assertion to be proved is then

$$A(a_1, \ldots, a_m, n) \rightarrow \Diamond B(a_1, \ldots, a_m, n)$$

where n is a free variable over $\mathbb{N}_0$ and $a_1, \ldots, a_m$ are all other free variables occurring in A and B. The proof of this latter formula runs by induction on n, i.e., according to the following proof rule:

$$A(a_1, \ldots, a_m, 0) \to \Diamond B(a_1, \ldots, a_m, 0),$$
$$A(a_1, \ldots, a_m, n) \to \Diamond A(f_1(a_1, \ldots, a_m, n), \ldots, f_m(a_1, \ldots, a_m, n), n-1),$$
$$B(f_1(a_1, \ldots, a_m, n), \ldots, f_m(a_1, \ldots, a_m, n), n-1) \to B(a_1, \ldots, a_m, n)$$
$$\vdash A(a_1, \ldots, a_m, n) \to \Diamond B(a_1, \ldots, a_m, n)$$

where $f_1, \ldots, f_m$ are appropriately chosen functions.

This proof rule is sound, since we may join the induction hypothesis

$$A(f_1(a_1, \ldots, a_m, n), \ldots, f_m(a_1, \ldots, a_m, n), n-1) \to$$
$$\Diamond B(f_1(a_1, \ldots, a_m, n), \ldots, f_m(a_1, \ldots, a_m, n), n-1)$$

to the hypotheses and deduce the conclusion by induction.

We call the method using just this proof rule (i.e., proving $A(a_1, \ldots, a_m, n) \to \Diamond B(a_1, \ldots, a_m, n)$ by "normal" mathematical induction) the *simple intermittent assertion method.* Instances of the general method are provided by the following two possible generalizations of this induction:

3) The induction may run with more complicated induction steps. For example, we need not directly have:

$$B(f_1(a_1, \ldots, a_m, n), \ldots, f_m(a_1, \ldots, a_m, n), n-1) \to B(a_1, \ldots, a_m, n).$$

It suffices to show:

$$B(f_1(a_1, \ldots, a_m, n), \ldots, f_m(a_1, \ldots, a_m, n), n-1) \to \Diamond B(a_1, \ldots, a_m, n)$$

for which we may, of course, again use the induction hypothesis. As an example, we could have:

$$B(f_1(a_1, \ldots, a_m, n), \ldots, f_m(a_1, \ldots, a_m, n), n-1) \to$$
$$\Diamond A(f_1'(a_1, \ldots, a_m, n), \ldots, f_m'(a_1, \ldots, a_m, n), n-1),$$

hence, applying the induction hypothesis

$$B(f_1(a_1, \ldots, a_m, n), \ldots, f_m(a_1, \ldots, a_m, n), n-1) \to$$
$$\Diamond B(f_1'(a_1, \ldots, a_m, n), \ldots, f_m'(a_1, \ldots, a_m, n), n-1)$$

and then, say,

$$B(f_1'(a_1, \ldots, a_m, n), \ldots, f_m'(a_1, \ldots, a_m, n), n-1) \to B(a_1, \ldots, a_m, n).$$

4) Instead of induction on $n \in \mathbb{N}_0$ any (transfinite) induction over a well-founded ordering is possible.

We dispense with a formulation of a general scheme of all these possible proof strategies as a single proof rule. We will give an example in the next section which runs in just the way indicated in 3).

We conclude this section by observing a relationship between the loop invariant:

$$P(a, b, n) \equiv a = 2^{n-b}$$

used for the partial correctness proof with Hoare's calculus in Section 24 and the intermittent assertion:

$$B(a_0, a, n) \equiv \text{at } \alpha_3 \wedge a = a_0 \times 2^n.$$

The essential part of B is $B'(a_0, a, n) \equiv a = a_0 \times 2^n$. If we denote by $A'(a)$ the "starting condition" $a=1$ of the loop, then the following relationship holds:

$$P(a, b, n) \leftrightarrow \exists y(A'(y) \wedge \forall x(B'(a, x, b) \rightarrow B'(y, x, n))).$$

In fact, the formula on the right side of $\leftrightarrow$ evaluates to:

$$\exists y(y=1 \wedge \forall x(x=a\times 2^b \rightarrow x=y\times 2^n))$$

which is equivalent to:

$$\exists y(y=1 \wedge a\times 2^b = y\times 2^n)$$

and hence to:

$$a\times 2^b = 2^n$$

which means $a=2^{n-b}$ and is just P.

The scheme of the above formula can be slightly extended to a general relationship between loop invariants and intermittent assertions (for the same loop) used in the simple intermittent assertion method.

26. Examples of Applications

In this last section we illustrate the two verification methods discussed in Sections 24 and 25 by giving a further example for each.

Hoare's calculus is so well established and is, as we have already mentioned, not necessarily bound to temporal logic of programs, so that a further example is essentially intended to show once more the use of the labels entry(ψ) and after(ψ) instead of the program piece ψ itself, which is the common way of using the calculus. Notice again that the use of labels reduces the amount of text which must be written down in the verification. We content ourselves with a simple standard example which has the advantage that it contains all sequential programming constructs of our language (except **if** ... **then** ... **fi**).

Program Example 26.1

```
ψ8 ≡ initial a=n ∧ b=m;
    α0: c:=1;
    α1: while b≠0 do
            α2: if odd(b) then α3: b:=b−1;
                               α4: c:=c×a
                          else α5: b:=b/2;
                               α6: a:=a×a fi
                    od;
    α7: stop
```

ψ_8 computes the value of n^m (for $n, m \in \mathbb{N}_0$ and defining $0^0=1$) on the variable c. We want to prove its partial correctness.

Assertion

$$\psi_8 \vdash a=n \wedge b=m[\alpha_0, \alpha_7]\, c=n^m.$$

Proof of the assertion. Let $P \equiv c \times a^b = n^m$. We first show that P is an invariant of the loop of ψ_8:

(1) $\quad P \wedge b \neq 0[\alpha_2, \alpha_1]\, P.$

Derivation of (1)

(1.1)	$P \wedge b \neq 0 \wedge \mathrm{odd}(b)[\alpha_3, \alpha_4]\, c \times a^{b+1} = n^m$	(Π), (H1)
(1.2)	$c \times a^{b+1} = n^m[\alpha_4, \alpha_1]\, P$	(Π), (H1)
(1.3)	$P \wedge b \neq 0 \wedge \mathrm{odd}(b)[\alpha_3, \alpha_1]\, P$	(H2), (1.1), (1.2)
(1.4)	$P \wedge b \neq 0 \wedge \neg\,\mathrm{odd}(b)[\alpha_5, \alpha_6]\, c \times a^{2 \times b} = n^m$	(Π), (H1)
(1.5)	$c \times a^{2 \times b} = n^m[\alpha_6, \alpha_1]\, P$	(Π), (H1)
(1.6)	$P \wedge b \neq 0 \wedge \neg\,\mathrm{odd}(b)[\alpha_5, \alpha_1]\, P$	(H2), (1.4), (1.5)
(1.7)	$P \wedge b \neq 0[\alpha_2, \alpha_1]\, P$	(H3), (1.3), (1.6)

The remaining derivation runs as follows:

(2)	$P[\alpha_1, \alpha_7]\, P \wedge b=0$	(H5), (1)
(3)	$P \wedge b=0 \rightarrow c=n^m$	(data)
(4)	$P[\alpha_1, \alpha_7]\, c=n^m$	(H6), (2), (3)
(5)	$a=n \wedge b=m[\alpha_0, \alpha_1]\, a=n \wedge b=m \wedge c=1$	(Π), (H1)
(6)	$a=n \wedge b=m \wedge c=1 \rightarrow P$	(data)
(7)	$a=n \wedge b=m[\alpha_0, \alpha_1]\, P$	(H6), (5), (6)
(8)	$a=n \wedge b=m[\alpha_0, \alpha_7]\, c=n^m$	(H2), (4), (7) □

We take as an example for the intermittent assertion method one with a more complicated induction mechanism than formulated in the simple version of the method. We assume $f_1, f_2, f_3, \ldots$ to be a sequence of unary function symbols and let g be another function symbol with the property:

$$g(n)=i \leftrightarrow \exists j(n=2^{i-1} \times (2j+1)) \quad (n, i, j \in \mathbb{N}_0, n>0).$$

g obviously determines for $n>0$ the value

1 + exponent of 2 in the prime factorization of n.

Program Example 26.2

$\psi_9 \equiv$ **initial** $a=m \wedge c=1$;
$\quad\alpha_0$: **while** $c \leq 2^n-1$ **do** $\alpha_1: i:=g(c)$;
$\quad\quad\quad\quad\quad\quad\quad\quad\quad\alpha_2: a:=f_i(a)$;
$\quad\quad\quad\quad\quad\quad\quad\quad\quad\alpha_3: c:=c+1$ **od**;
$\quad\alpha_4$: **stop**

This program computes the following recursively defined function h on the variable a:

$$h(n, m) = \texttt{if}\, n=0\ \texttt{then}\ m\ \texttt{else}\, h(n-1, f_n(h(n-1, m)))\ \texttt{fi}$$

where $n, m \in \mathbb{N}_0$.

Assertion

$$\psi_9 \vdash \text{start}_{\psi_9} \rightarrow \Diamond(\text{at } \alpha_4 \wedge a = h(n, m)).$$

Proof of the assertion. Let a_0, k, l be further variables (over $\mathbb{N}_0$) and

$$A(a, c, n, a_0, k, l) \equiv \text{at } \alpha_0 \wedge a = a_0 \wedge c = k \times 2^l + 1 \wedge c \leq 2^n,$$
$$B(a, c, n, a_0, k, l) \equiv \text{at } \alpha_0 \wedge a = h(l, a_0) \wedge c = (k+1) \times 2^l \wedge c \leq 2^n.$$

The essential part of our proof is to show:

(1) $A(a, c, n, a_0, k, l) \rightarrow \Diamond B(a, c, n, a_0, k, l)$ for $l \leq n$.

This is done by induction on l.

Derivation of (1) *for* $l=0$

(1.1)	$a_0 = h(0, a_0)$	def. of h
(1.2)	$\text{at } \alpha_0 \wedge a = a_0 \wedge c = k \times 2^0 + 1 \wedge c \leq 2^n \rightarrow$ $\text{at } \alpha_0 \wedge a = h(0, a_0) \wedge c = (k+1) \times 2^0 \wedge c \leq 2^n$	(data)
(1.3)	$A(a, c, n, a_0, k, 0) \rightarrow B(a, c, n, a_0, k, 0)$	(1.2)
(1.4)	$A(a, c, n, a_0, k, 0) \rightarrow \Diamond B(a, c, n, a_0, k, 0)$	(1.3)

Derivation of (1) *for* $l>0$ *(and* $l \leq n$*).*

(1.5)	$c = k \times 2^l + 1 \rightarrow c = (2 \times k) \times 2^{l-1} + 1$	(data)
(1.6)	$A(a, c, n, a_0, k, l) \rightarrow A(a, c, n, a_0, 2 \times k, l-1)$	(1.5)
(1.7)	$A(a, c, n, a_0, 2 \times k, l-1) \rightarrow \Diamond B(a, c, n, a_0, 2 \times k, l-1)$	ind. hyp.
(1.8)	$c = (2k+1) \times 2^{l-1} \wedge c \leq 2^n \rightarrow c \leq 2^n - 1$	(data)
(1.9)	$B(a, c, n, a_0, 2 \times k, l-1) \rightarrow$ $\circ(\text{at } \alpha_1 \wedge a = h(l-1, a_0) \wedge c = (2 \times k+1) \times 2^{l-1} \wedge c \leq 2^n - 1)$	(Π), (1.8)
(1.10)	$c = (2 \times k+1) \times 2^{l-1} \rightarrow g(c) = l$	def. of g
(1.11)	$\text{at } \alpha_1 \wedge a = h(l-1, a_0) \wedge c = (2 \times k+1) \times 2^{l-1} \wedge c \leq 2^n - 1 \rightarrow$ $\circ(\text{at } \alpha_2 \wedge a = h(l-1, a_0) \wedge$ $c = (2 \times k+1) \times 2^{l-1} \wedge c \leq 2^n - 1 \wedge i = l)$	(Π), (1.10)
(1.12)	$\text{at } \alpha_2 \wedge a = h(l-1, a_0) \wedge c = (2 \times k+1) \times 2^{l-1} \wedge c \leq 2^n - 1 \wedge i = l \rightarrow$ $\circ(\text{at } \alpha_3 \wedge a = f_l(h(l-1, a_0)) \wedge c = (2 \times k+1) \times 2^{l-1} \wedge c \leq 2^n - 1)$	(Π)
(1.13)	$\text{at } \alpha_3 \wedge a = f_l(h(l-1, a_0)) \wedge c = (2 \times k+1) \times 2^{l-1} \wedge c \leq 2^n - 1 \rightarrow$ $\circ(\text{at } \alpha_0 \wedge a = f_l(h(l-1, a_0)) \wedge c = (2 \times k+1) \times 2^{l-1} + 1 \wedge c \leq 2^n)$	(Π)
(1.14)	$A(a, c, n, a_0, 2 \times k, l-1) \rightarrow$ $\Diamond A(a, c, n, f_l(h(l-1, a_0)), 2 \times k+1, l-1)$	(1.7), (1.9), (1.10)–(1.13)
(1.15)	$A(a, c, n, f_l(h(l-1, a_0)), 2 \times k+1, l-1) \rightarrow$ $\Diamond(\text{at } \alpha_0 \wedge a = h(l-1, f_l(h(l-1, a_0))) \wedge$ $c = (2 \times k+2) \times 2^{l-1} \wedge c \leq 2^n)$	ind. hyp.

(1.16)	$h(l-1, f_l(h(l-1, a_0))) = h(l, a_0)$	def. of h
(1.17)	$(2\times k+2)\times 2^{l-1} = (k+1)\times 2^l$	(data)
(1.18)	$A(a, c, n, a_0, k, l) \to \Diamond B(a, c, n, a_0, k, l)$	(1.6), (1.14)–(1.17)

Now taking $a_0 = m$, $k = 0$, and $l = n$ in (1) we find:

(2) $\quad \text{at } \alpha_0 \wedge a = m \wedge c = 1 \wedge c \le 2^n \to \Diamond(\text{at } \alpha_0 \wedge a = h(n, m) \wedge c = 2^n \wedge c \le 2^n)$

and the rest is quite easy:

(3)	$\text{start}_{\psi_9} \to \text{at } \alpha_0 \wedge a = m \wedge c = 1 \wedge c \le 2^n$	(data)
(4)	$\text{at } \alpha_0 \wedge a = h(n, m) \wedge c = 2^n \to \circ(\text{at } \alpha_4 \wedge a = h(n, m))$	(Π)
(5)	$\text{start}_{\psi_9} \to \Diamond(\text{at } \alpha_4 \wedge a = h(n, m))$	(2), (3), (4) □

Bibliographical Remarks

Chapter I

The atnext operator chosen as a basic operator in our presentation was introduced in Kröger (1984b). The until operator was already known from Kamp (1968), introduced into the context of program analysis by Gabbay et al. (1980) and subsequently used as a strong and weak (unless) operator. The before operator can be found in Manna and Pnueli (1982b) and the while operator in Lamport (1980). The latter also gives some arguments against the use of the nexttime operator, and see also Lamport (1983b). Suggestions for more expressive temporal languages are presented in Wolper (1983). Our semantical apparatus is an obvious application of the usual "possible world" semantics of modal logic as developed by Kripke (1963). Slightly different semantics are compared in Emerson (1983). Other extensive lists of temporal logical laws also containing some not presented here can be found in Hailpern (1982) and Manna and Pnueli (1982b, 1983c).

Chapter II

The formal system Σ_{TA} seems to be minimal with respect to the number of ("independent") axioms and rules. It was first developed in Schlingloff (1983). Pnueli's system for a logic with `until` instead of `atnext` can be found in Gabbay et al. (1980). Our completeness proof tries to remain as close as possible to the classical so-called Henkin-Hasenjäger method. A first similar proof (for another logic) was given in Kröger (1977). A slightly different proof for a branching time temporal logic is contained in Ben-Ari et al. (1983). The induction principle for the atnext operator is introduced in Kröger (1984a). The induction principle for the unless operator can also be found in Wolper (1983).

Chapter III

A formal proof system for first-order temporal logic was given in Manna and Pnueli (1979) which, however, is (probably) too weak. A system analogous to Σ_{TP} can be found in Manna and Pnueli (1983c). The accentuation of the principle of well-founded orderings which was present in termination and total correctness proofs of programs for a long time as a purely temporal logic proof principle is performed in Manna and Pnueli (1983a) and Kröger (1984a, 1985a).

Chapter IV

The class of programs investigated in this book is essentially the same as in Manna and Pnueli (1982b) with the main difference being that we have taken structured programs instead of transition graphs (i.e., goto's). A somewhat larger class is treated in Owicki and Lamport (1982). The program axioms are an adjusted version of those in Kröger (1984a, 1985a). A first "translation" of parallel programs into temporal formulas, i.e., a temporal semantics of parallel programs was given by Pnueli (1979, 1981). A somewhat different approach to the specification of a program is in Owicki and Lamport (1982). The use of formulas of the kind α besides those of the kind at α was suggested in Kröger (1983, 1984a, 1985a). The classification and choice of program properties follows Manna and Pnueli (1982b). The use of the atnext operator is from Kröger (1984a, b, 1985a).

Chapter V

The general idea of elaborating "high-level" program verification principles is addressed in Manna and Pnueli (1983a). The basic invariant method underlies – in one form or another – every comparable approach to verification of safety properties. Methods for "constructing" appropriate invariants are discussed in Manna and Pnueli (1982a). The proof rule (atnext) and the rules for the other operators are from Kröger (1984a, 1985a). In Kröger (1984b) some other proof rules for **atnext** are discussed. Manna and Pnueli (1983b, c) give somewhat different rules for the unless operator. Example 17.3 is also dealt with in Manna and Pnueli (1982a) where, however, other properties of the program are proved. The alternating bit protocol (Example 19.1) was introduced by Bartlett et al. (1969). A detailed discussion and a different proof can be found in Hailpern (1982) and Hailpern and Owicki (1983). The algorithm of Peterson (1981) in Example 19.3 is also treated in Manna and Pnueli (1983b) using the unless operator.

Chapter VI

The investigation of fairness, justice and impartiality is due to Lehmann et al. (1981). Further notions of fairness for a CSP-like language can be found in Kuiper and de Roever (1983). The proof rules for liveness properties presented in this book are slight modifications of rules developed in Manna and Pnueli (1982a, c). A similar approach is in Owicki and Lamport (1982). Moreover, graphical representations of liveness proofs are investigated in these latter papers. Program Example 23.2 is due to Manna and Pnueli (1982a) where a different termination proof is given. For the alternating bit protocol in Example 23.3 compare again Hailpern (1982) and Hailpern and Owicki (1983).

Chapter VII

An extensive summary of Hoare's (1969) calculus is given by Apt (1981). The approach was extended to parallel programs by Owicki and Gries (1976). The intermittent assertion method was introduced by Burstall (1974) and further developed by Manna and Waldinger (1978). Another formal representation within temporal logic is given by Apt and Delporte (1983). Relationships between loop invariants and intermittent assertions are given in Kröger (1980). The program of Example 26.2 is due to Partsch and Pepper (1976).

Appendix

Table of Laws and Rules

Axioms and rules of Σ_{TA} (p. 25)

(taut)	All tautologically valid formulas
(ax 1)	$\neg \circ A \leftrightarrow \circ \neg A$
(ax 2)	$\circ(A \to B) \to (\circ A \to \circ B)$
(ax 3)	$\Box A \to A \wedge \circ \Box A$
(ax 4)	$\circ \Box \neg B \to A \; \mathbf{atnext} \; B$
(ax 5)	$A \; \mathbf{atnext} \; B \leftrightarrow \circ(B \to A) \wedge \circ(\neg B \to A \; \mathbf{atnext} \; B)$
(mp)	$A, A \to B \vdash B$
(nex)	$A \vdash \circ A$
(ind)	$A \to B, \; A \to \circ A \vdash A \to \Box B$

Additional axioms and rules of Σ_{TP} (p. 45, 46, 53)

(ax 6)	$\forall x A \to A_x(t)$ if t is substitutable for x in A
(ax 7)	$\forall x \circ A \to \circ \forall x A$
(ax 8)	$A \to \circ A$ if A does not contain local variables
(eq 1)	$x = x$
(eq 2)	$x = y \to (A \to A_x(y))$ if A does not contain temporal operators
(gen)	$A \to B \vdash A \to \forall x B$ if there is no free occurrence of x in A

[Modification of (ax 8) in $\mathscr{L}^{+}_{TP}$:
(ax 8^+) $A \to \circ A$ if A contains neither local variables nor propositional variables]

Additional axioms for languages with well-founded ordering (p. 50)

(po 1)	$z \preccurlyeq z$
(po 2)	$z_1 \preccurlyeq z_2 \wedge z_2 \preccurlyeq z_3 \to z_1 \preccurlyeq z_3$
(po 3)	$z_1 \preccurlyeq z_2 \wedge z_2 \preccurlyeq z_1 \to z_1 = z_2$
(ti)	$\forall z(\forall z'(z' \prec z \to A(z')) \to A(z)) \to A(z)$

Further logical laws and rules (p. 17, 18, 19, 20, 21, 22, 26, 27, 37, 40, 46, 47, 48)

(T 1)	$\neg \circ A \leftrightarrow \circ \neg A$
(T 2)	$\neg \Box A \leftrightarrow \Diamond \neg A$
(T 3)	$\neg \Diamond A \leftrightarrow \Box \neg A$
(T 4)	$\Box A \to A$
(T 5)	$A \to \Diamond A$
(T 6)	$\Box A \to \circ A$
(T 7)	$\circ A \to \Diamond A$

(T8) $\Box A \to \Diamond A$
(T9) $\Box A \to A$ **atnext** B
(T10) $\Diamond\Box A \to \Box\Diamond A$
(T11) $\circ A \leftrightarrow A$ **atnext true**
(T12) $\Box A \leftrightarrow A \wedge$ **false atnext** $\neg A$
(T13) $\Diamond A \leftrightarrow A \vee \neg$(**false atnext** A)
(T14) $\Box\Box A \leftrightarrow \Box A$
(T15) $\Diamond\Diamond A \leftrightarrow \Diamond A$
(T16) $\Box\circ A \leftrightarrow \circ\Box A$
(T17) $\Diamond\circ A \leftrightarrow \circ\Diamond A$
(T18) $\circ(A \to B) \leftrightarrow \circ A \to \circ B$
(T19) $\circ(A \wedge B) \leftrightarrow \circ A \wedge \circ B$
(T20) $\circ(A \vee B) \leftrightarrow \circ A \vee \circ B$
(T21) $\circ(A$ **atnext** $B) \leftrightarrow \circ A$ **atnext** $\circ B$
(T22) $\Box(A \wedge B) \leftrightarrow \Box A \wedge \Box B$
(T23) $\Diamond(A \vee B) \leftrightarrow \Diamond A \vee \Diamond B$
(T24) $(A \wedge B)$ **atnext** $C \leftrightarrow A$ **atnext** $C \wedge B$ **atnext** C
(T25) $(A \vee B)$ **atnext** $C \leftrightarrow A$ **atnext** $C \vee B$ **atnext** C
(T26) $\Box(A \to B) \to (\Box A \to \Box B)$
(T27) $\Box A \vee \Box B \to \Box(A \vee B)$
(T28) $(\Diamond A \to \Diamond B) \to \Diamond(A \to B)$
(T29) $\Diamond(A \wedge B) \to \Diamond A \wedge \Diamond B$
(T30) A **atnext** $(B \vee C) \to A$ **atnext** $B \vee A$ **atnext** C
(T31) $\Box A \leftrightarrow A \wedge \circ\Box A$
(T32) $\Diamond A \leftrightarrow A \vee \circ\Diamond A$
(T33) A **atnext** $B \leftrightarrow \circ(B \to A) \wedge \circ(\neg B \to A$ **atnext** $B)$
(T34) $A \to B \vdash \circ A \to \circ B$
(T35) $A \to B \vdash \Diamond A \to \Diamond B$
(T36) $A \to B \vdash A$ **atnext** $C \to B$ **atnext** C
(T37) $A \vdash \circ B \to \circ(A \wedge B)$
(T38) $A \vdash \Box B \to \Box(A \wedge B)$
(T39) $A \vdash \Diamond B \to \Diamond(A \wedge B)$
(T40) $A \vdash B$ **atnext** $C \to (A \wedge B)$ **atnext** $(A \wedge C)$
(T41) $\Box A \to B \vdash \Box A \to \Box B$
(T42) $A \to \Diamond B \vdash \Diamond A \to \Diamond B$
(T43) A **until** $B \leftrightarrow B$ **atnext** $(A \to B) \wedge \circ\Diamond B$
(T44) A **unless** $B \leftrightarrow B$ **atnext** $(A \to B)$
(T45) A **while** $B \leftrightarrow \neg B$ **atnext** $(A \to \neg B)$
(T46) A **before** $B \leftrightarrow \neg B$ **atnext** $(A \vee B)$
(T47) $\Box\Diamond(A \vee B) \leftrightarrow \Box\Diamond A \vee \Box\Diamond B$
(T48) $B \to A \vdash A$ **atnext** B
(T49) $\forall x \circ A \leftrightarrow \circ \forall x A$
(T50) $\exists x \circ A \leftrightarrow \circ \exists x A$
(T51) $\forall x \Box A \leftrightarrow \Box \forall x A$
(T52) $\exists x \Diamond A \leftrightarrow \Diamond \exists x A$
(T53) $\forall x(A$ **atnext** $B) \leftrightarrow (\forall x A)$ **atnext** B
if there is no free occurrence of x in B

(T54) $\exists x(A\ \mathtt{atnext}\ B) \leftrightarrow (\exists x A)\ \mathtt{atnext}\ B$
if there is no free occurrence of x in B

(alw) $A \vdash \Box A$

(som) $A \to \circ B \vdash A \to \Diamond B$

(chain) $A \to \Diamond B,\ B \to \Diamond C \vdash A \to \Diamond C$

(prop) $A_1, \ldots, A_n \vdash B$ if B is a tautological consequence of $A_1, \ldots, A_n$

(pred) $A_1, \ldots, A_n \vdash B$ if B is a first-order consequence of $A_1, \ldots, A_n$

Logical induction principles (p. 38, 39, 40, 41)

(ind) $A \to B,\ A \to \circ A \vdash A \to \Box B$

(ind′) $A \to \circ A \vdash A \to \Box A$

(ind″) $A \to B,\ B \to \circ B \vdash A \to \Box B$

(gind) $A \to C\ \mathtt{atnext}\ B,\ B \wedge C \to C\ \mathtt{atnext}\ B \vdash A \to \Box\circ(B \to C)$

(indatnext) $A \to \circ(C \to B) \wedge \circ(\neg C \to A) \vdash A \to B\ \mathtt{atnext}\ C$

(indatnext^n) $A \to \circ(C \to B_1) \wedge \circ(\neg C \to A)$,
$B_1 \to \circ(C \to B_2) \wedge \circ(\neg C \to B_1)$,
$\vdots$
$B_{n-1} \to \circ(C \to B) \wedge \circ(\neg C \to B_{n-1})$
$\vdash A \to B\ \mathtt{atnext}^n\ C$

(indunless) $A \to \circ C \vee \circ(A \wedge B) \vdash A \to B\ \mathtt{unless}\ C$

(indwhile) $A \to \circ(C \to A \wedge B) \vdash A \to B\ \mathtt{while}\ C$

(indbefore) $A \to \circ\neg C \wedge \circ(A \vee B) \vdash A \to B\ \mathtt{before}\ C$

Well-founded ordering rule (p. 50)

(wfo) $A(z) \to \Diamond(B \vee \exists z'(z' \prec z \wedge A(z'))) \vdash \exists z A(z) \to \Diamond B$
if B does not contain z

Program axioms (and rule) (p. 64, 65, 66, 67, 104, 105, 106)

(B1) $\mathrm{start}_\Pi \to \Box A \vdash A$

(B2) $\mathrm{nil}_\Pi \wedge A \to \circ(\mathrm{nil}_\Pi \wedge A)$

(B3) $\Box\Diamond(\mathrm{at}\ \lambda \wedge E_\lambda) \to \Diamond\lambda$

(B3′) $\Box\Diamond E^{(i)} \to \Diamond(\alpha_0^{(i)} \vee \ldots \vee \alpha_m^{(i)})$ for $i = 1, \ldots, p$

(J) $\Diamond\Box(\mathrm{at}\ \lambda \wedge E_\lambda) \to \Diamond\lambda$

(J′) $\Diamond\Box E^{(i)} \to \Diamond(\alpha_0^{(i)} \vee \ldots \vee \alpha_m^{(i)})$ for $i = 1, \ldots, p$

(Imp) $\Diamond(\alpha_0^{(i)} \vee \ldots \vee \alpha_m^{(i)})$ for $i = 1, \ldots, p$

(Π1) $\lambda \to \neg\lambda'$ for $\lambda \neq \lambda'$

(Π2) $\lambda \to \mathrm{at}\ \lambda$

(Π3) $\mathrm{at}\ \alpha_j^{(i)} \to \neg\,\mathrm{at}\ \alpha_k^{(i)}$ for $j \neq k,\ j, k = 1, \ldots, m_i,\ i = 1, \ldots, p$

(Π4) $\mathrm{at}\ \lambda \wedge E_\lambda \to \neg\,\mathrm{nil}_\Pi$

(Π5) $\mathrm{at}\ \lambda \wedge \neg\lambda \to \circ\ \mathrm{at}\ \lambda$

(Π6) $\mathrm{at}\ \alpha_e^{(i)} \to \circ\ \mathrm{at}\ \alpha_e^{(i)}$ for $i = 1, \ldots, p$

(Π7) $\lambda \wedge P \to \circ P$ if λ is the label of a statement $\notin \mathcal{A}$ and P is a P-formula

(CS) $\lambda \to (C_1 \wedge \circ\ \mathrm{at}\ \lambda_1) \vee \ldots \vee (C_q \wedge \circ\ \mathrm{at}\ \lambda_q)$
where $(\lambda, C_1, \lambda_1), \ldots, (\lambda, C_q, \lambda_q) \in \mathrm{trans}(\Pi_i)$ for some $i = 1, \ldots, p$,
and $\mathrm{trans}(\Pi_i)$ contains no other element of the form $(\lambda, \ldots)$

(data) Every formula from $\mathrm{Th}(\mathbf{S}_\Pi)$

(assign) $\lambda \wedge P_a(t) \rightarrow \circ P$ (P P-formula)
for every statement λ: $a := t$ or λ: **await** B **then** $a := t$ occurring in Π

Program verification rules (p. 70, 79, 80, 90, 91, 92, 107, 108, 112, 121)

(trans) $\alpha \wedge A \rightarrow \circ B$ for every $\alpha \in \bar{\mathcal{M}}_\Pi$,
$\text{nil}_\Pi \wedge A \rightarrow B$
$\vdash A \rightarrow \circ B$

(inv) $A \rightarrow B$, B **invof** $\bar{\mathcal{M}}_\Pi \vdash A \rightarrow \Box B$

(inv′) $\text{start}_\Pi \rightarrow A$, A **invof** $\bar{\mathcal{M}}_\Pi \vdash \Box A$

(atnext) $\alpha \wedge A \rightarrow \circ(C \rightarrow B) \wedge \circ(\neg C \rightarrow A)$ for every $\alpha \in \bar{\mathcal{M}}_\Pi$,
$\text{nil}_\Pi \wedge A \rightarrow (C \rightarrow B)$
$\vdash A \rightarrow B$ **atnext** C

(atnext²) $\alpha \wedge A \rightarrow \circ(C \rightarrow B_1) \wedge \circ(\neg C \rightarrow A)$ for every $\alpha \in \bar{\mathcal{M}}_\Pi$,
$\text{nil}_\Pi \wedge A \rightarrow (C \rightarrow B_1)$,
$\alpha \wedge B_1 \rightarrow \circ(C \rightarrow B) \wedge \circ(\neg C \rightarrow B_1)$ for every $\alpha \in \bar{\mathcal{M}}_\Pi$,
$\text{nil}_\Pi \wedge B_1 \rightarrow (C \rightarrow B)$
$\vdash A \rightarrow B$ **atnext**^{2} C

(disatnext^n) $\alpha \wedge A \rightarrow \circ(C \rightarrow B_1) \wedge \circ(\neg C \rightarrow A)$ for every $\alpha \in \bar{\mathcal{M}}_\Pi$,
$\alpha \wedge B_1 \rightarrow \circ(C \rightarrow B_2) \wedge \circ(\neg C \rightarrow B_1)$ for every $\alpha \in \bar{\mathcal{M}}_\Pi$,
⋮
$\alpha \wedge B_{n-1} \rightarrow \circ(C \rightarrow B) \wedge \circ(\neg C \rightarrow B_{n-1})$ for every $\alpha \in \bar{\mathcal{M}}_\Pi$,
$\text{nil}_\Pi \wedge A \rightarrow (C \rightarrow B_1)$,
$\text{nil}_\Pi \wedge B_1 \rightarrow (C \rightarrow B_2)$,
⋮
$\text{nil}_\Pi \wedge B_{n-1} \rightarrow (C \rightarrow B)$
$\vdash A \vee B_1 \vee \ldots \vee B_{n-1} \rightarrow B$ **atnext** $C \vee \ldots \vee B$ **atnext**^{n} C

(unless) $\alpha \wedge A \rightarrow \circ C \vee \circ(A \wedge B)$ for every $\alpha \in \bar{\mathcal{M}}_\Pi$,
$\text{nil}_\Pi \wedge A \rightarrow B \vee C$
$\vdash A \rightarrow B$ **unless** C

(while) $\alpha \wedge A \rightarrow \circ(C \rightarrow A \wedge B)$ for every $\alpha \in \bar{\mathcal{M}}_\Pi$,
$\text{nil}_\Pi \wedge A \rightarrow (C \rightarrow B)$
$\vdash A \rightarrow B$ **while** C

(before) $\alpha \wedge A \rightarrow \circ\neg C \wedge \circ(A \vee B)$ for every $\alpha \in \bar{\mathcal{M}}_\Pi$,
$\text{nil}_\Pi \wedge A \rightarrow \neg C$
$\vdash A \rightarrow B$ **before** C

(event0) A **invof** $\bar{\mathcal{M}}_\Pi \setminus \{\alpha\}$,
$\alpha \wedge A \rightarrow \circ B$,
$\Box A \rightarrow \Diamond(\text{at } \alpha \wedge E_\alpha)$
(for arbitrary $\alpha \in \bar{\mathcal{M}}_\Pi$)
$\vdash A \rightarrow \Diamond B$

(event1) $\alpha \wedge \text{at } \beta_i \wedge A \rightarrow \circ A$ for every $\alpha \in \bar{\mathcal{M}}_\Pi \setminus \bar{\mathcal{M}}_{\Pi_i}$,
$\beta_i \wedge A \rightarrow \circ B$,
$\Box(\text{at } \beta_i \wedge A) \rightarrow \Diamond E_{\beta_i}$
$\vdash \text{at } \beta_i \wedge A \rightarrow \Diamond B$

(event2) $\beta_i \to \circ B,\ \square\, \text{at}\ \beta_i \to \Diamond E_{\beta_i} \vdash \text{at}\ \beta_i \to \Diamond B$

(well) $\alpha \wedge A(z) \to \circ(B \vee \exists z'(z' \preccurlyeq z \wedge A(z')))$ for every $\alpha \in \bar{\mathscr{M}}_{\Pi} \setminus \bar{\mathscr{M}}_{\Pi_{h(z)}}$,
$\alpha \wedge A(z) \to \circ(B \vee \exists z'(z' \preccurlyeq z \wedge A(z')))$ for every $\alpha \in \bar{\mathscr{M}}_{\Pi_{h(z)}}$,
$\square A(z) \to \Diamond(B \vee E^{(h(z))})$
$\vdash \exists z A(z) \to \Diamond B$
(z not contained in B)

(HO) $\text{start}_{\Pi} \wedge P \to Q$ **atnext** $\text{at}\ \alpha_e \vdash \text{start}_{\Pi} \wedge P \to \square(\text{at}\ \alpha_e \to Q)$

Hoare's calculus (p. 123)

(H1) $\beta_1 \wedge P \to \circ(\text{at}\ \beta_2 \wedge Q) \vdash P[\beta_1, \beta_2]Q$ if ψ is an elementary statement

(H2) $P[\beta_1, \gamma]P',\ P'[\gamma, \beta_2]Q \vdash P[\beta_1, \beta_2]Q$
if $\psi \equiv \psi_1 ; \psi_2$, $\gamma = \text{after}(\psi_1) = \text{entry}(\psi_2)$

(H3) $P \wedge B[\gamma_1, \beta_2]Q,\ P \wedge \neg B[\gamma_2, \beta_2]Q \vdash P[\beta_1, \beta_2]Q$
if $\psi \equiv$ **if** B **then** ψ_1 **else** ψ_2 **fi**, $\gamma_1 = \text{entry}(\psi_1)$, $\gamma_2 = \text{entry}(\psi_2)$

(H4) $P \wedge B[\gamma_1, \beta_2]Q,\ P \wedge \neg B \to Q \vdash P[\beta_1, \beta_2]Q$
if $\psi \equiv$ **if** B **then** ψ_1 **fi**, $\gamma_1 = \text{entry}(\psi_1)$

(H5) $P \wedge B[\gamma_1, \beta_1]P \vdash P[\beta_1, \beta_2]P \wedge \neg B$
if $\psi \equiv$ **while** B **do** ψ_1 **od**, $\gamma_1 = \text{entry}(\psi_1)$

(H6) $P[\beta_1, \beta_2]P',\ P' \to Q \vdash P[\beta_1, \beta_2]Q$

(H7) $P \to P',\ P'[\beta_1, \beta_2]Q \vdash P[\beta_1, \beta_2]Q$

References

Sources

Apt, K.R. (1981): Ten years of Hoare's logic, a survey, part I. ACM Trans Program. Lang. Syst. *3*, 431–483

Apt, K.R., Delporte, C. (1983): An axiomatization of the intermittent assertion method using temporal logic. In: Diaz, J. (ed.): Automata, languages and programming, 10th Colloq., Barcelona (ICALP83). Lect. Notes Comput. Sci. *154*. Berlin-Heidelberg-New York-Tokyo: Springer, pp. 15–27

Bartlett, K.A., Scantlebury, R.A., Wilkinson, P.T. (1969): A note on reliable full-duplex transmission over half-duplex links. Commun. ACM *12*, 260–261

Ben-Ari, M. (1982a): Principles of concurrent programming. Englewood Cliffs: Prentice-Hall

Ben-Ari, M. (1982b): Temporal logic proofs of concurrent programs. Weizmann Inst. of Science, Rehovot, Israel, Rep. CS82-12

Ben-Ari, M., Manna, Z., Pnueli, A. (1983): The temporal logic of branching time. Acta Inf. *20*, 207–226

Burgess, J. (1982): Axioms for tense logic: 1. 'Since' and 'Until'. Notre Dame J. Formal Logic *23*, 367–374

Burstall, M. (1974): Program proving as hand simulation with a little induction. Proc. IFIP Congr. 1974, Stockholm. Amsterdam: North-Holland, pp. 308–312

Clifford, J. (1966): Tense logic and the logic of change. Logique Anal. *34*, 219–230

Emerson, E.A. (1983): Alternative semantics for temporal logics. Theor. Comput. Sci. *26*, 121–130

Gabbay, D., Pnueli, A., Shelah, S., Stavi, J. (1980): On the temporal analysis of fairness. Proc. 7th Ann. ACM Symp. Principles of Programming Languages, Las Vegas, NV, pp. 163–173

Hailpern, B.T. (1982): Verifying concurrent processes using temporal logic. Lect. Notes Comput. Sci. *129*. Berlin-Heidelberg-New York-Tokyo: Springer

Hailpern, B.T., Owicki, S.S. (1983): Modular verification of computer communication protocols. IEEE Trans. Commun. Com-*31*, 56–67

Hoare, C.A.R. (1969): An axiomatic basis for computer programming. Commun. ACM *12*, 576–580

Hughes, G.E., Cresswell, M.J. (1968): An introduction to modal logic. London: Methuen

Kamp, H.W. (1968): Tense logic and the theory of linear order. UCLA, Los Angeles, Ph.D. thesis

Kripke, S.A. (1963): Semantical analysis of modal logic I. Z. Math. Logik Grundlagen Math. *9*, 67–96

Kröger, F. (1975): Formalization of algorithmic reasoning. In: Bečvář, J. (ed.): Mathematical foundations of computer science 1975, 4th Symp., Mariánske Lázně. Lect. Notes Comput. Sci. *32*. Berlin-Heidelberg-New York-Tokyo: Springer, pp. 287–293

Kröger, F. (1976): Logical rules of natural reasoning about programs. ICALP 76, Edinburgh. Edinburgh: Edinburgh University Press, pp. 87–98

Kröger, F. (1977): LAR: A logic of algorithmic reasoning. Acta Inf. *8*, 243–266

Kröger, F. (1978): A uniform logical basis for the description, specification and verification of programs. Proc. IFIP Work. Conf. Formal Description of Programming Concepts, St. Andrews, Canada, 1977. Amsterdam: North-Holland, pp. 441–457

Kröger, F. (1980): Relationships between intermittent and invariant assertions. Techn. Univ. Munich, Inst. for Informatics, Rep. TUM-I8003

Kröger, F. (1983): Some new aspects of the temporal logic of concurrent programs. Techn. Univ. Munich, Inst. for Informatics, Rep. TUM-I8311

Kröger, F. (1984a): On the formal description and derivation of temporal proof rules for program properties. Techn. Univ. Munich, Inst. for Informatics, Rep. TUM-I8405

Kröger, F. (1984b): A generalized nexttime operator in temporal logic. J. Comput. Syst. Sci. *29*, 80–98

Kröger, F. (1985a): On temporal program verification rules. R.A.I.R.O Informatique théorique/Theoretical Informatics *19*, 261–280

Kröger, F. (1985b): Temporal logic of programs – Lecture notes. Techn. Univ. Munich, Inst. for Informatics, Rep. TUM-I8521

Kuiper, R., de Roever, W.P. (1983): Fairness assumptions for CSP in a temporal logic framework. Proc. IFIP Work. Conf. Formal Description of Programming Concepts II, Garmisch-Partenkirchen, 1982. Amsterdam: North-Holland, pp. 159–167

Lamport, L. (1980): 'Sometime' is sometimes 'Not Never': On the temporal logic of programs. Proc. 7th Ann. ACM Symp. Principles of Programming Languages, Las Vegas, NV, pp 174–185

Lamport, L. (1983a): Specifying concurrent program modules. ACM Trans. Program. Lang. Syst. *5*, 190–222

Lamport, L. (1983b): What good is temporal logic? Proc. IFIP Congress, Paris 1983. Amsterdam: North-Holland, pp 657–668

Lehmann, D., Pnueli, A., Stavi, J. (1981): Impartiality, justice and fairness: the ethics of concurrent termination. In: Even, S., Kariv, O. (eds.): Automata, languages and programming, 8th Colloq., Acre (ICALP81). Lect. Notes Comput. Sci. *115*. Berlin-Heidelberg-New York-Tokyo: Springer, pp. 264–277

Lemmon, E.J. (1966): An introduction to modal logic. Draft for a publication of Lemmon, E.J., Scott, D.: Intensional logic. K. Segerberg (ed.). Oxford: Blackwell, 1977

Manna, Z. (1980): Logics of programs. Proc. IFIP Congr. 1980, Tokyo. Amsterdam: North-Holland, pp. 41–51

Manna, Z., Pnueli, A. (1979): The modal logic of programs. In: Maurer, H.A. (ed.): Automata, languages and programming, 6th Colloq., Graz (ICALP79). Lect. Notes Comput. Sci. *71*. Berlin-Heidelberg-New York-Tokyo: Springer, pp. 385–409

Manna, Z., Pnueli, A. (1982a): Verification of concurrent programs: Temporal proof principles. In: Kozen, D. (ed.): Logics of programs, Workshop, Yorktown Heights. Lect. Notes Comput. Sci *131*. Berlin-Heidelberg-New York-Tokyo: Springer, pp. 200–252

Manna, Z., Pnueli, A. (1982b): Verification of concurrent programs: The temporal framework. In: Boyer, R.S., Moore, J.S. (eds.): The correctness problem in computer science. London: Academic Press, pp. 215–273

Manna, Z., Pnueli, A. (1982c): Verification of concurrent programs: Proving eventualities by well-founded ranking. Weizmann Inst. of Science, Rehovot, Israel, Rep. CS82-10

Manna, Z., Pnueli, A. (1983a): How to cook a temporal proof system for your pet language. Proc. 10th Ann. ACM Symp. Principles of Programming Languages, Austin, Texas, pp. 141–154

Manna, Z., Pnueli, A. (1983b): Proving precedence properties: The temporal way. In: Diaz, J. (ed.): Automata, languages and programming, 10th Colloq., Barcelona (ICALP83). Lect. Notes Comput. Sci. *154*. Berlin-Heidelberg-New York-Tokyo: Springer, pp. 491–512

Manna, Z., Pnueli, A.: (1983c). Verification of concurrent programs: A temporal proof system. In: Foundations of computer science IV. Amsterdam: Mathematical Centre Tracts *159*, pp. 163–255

Manna, Z., Waldinger, R. (1978): Is 'sometime' sometimes better than 'always'? Commun. ACM *21*, 159–172

Owicki, S.S., Gries, D. (1976): An axiomatic proof technique for parallel programs I. Acta Inf. *6*, 319–340

Owicki, S.S., Lamport, L. (1982): Proving liveness properties of concurrent programs. ACM Trans. Program. Lang. Syst. *4*, 455–495

Partsch, H., Pepper, P. (1976): A family of rules for recursion removal related to the Towers of Hanoi problem. Inf. Process. Lett. *5*, 174–177

Peterson, G.L. (1981): Myths about the mutual exclusion problem. Inf. Process. Lett. *12*, 115–116

Pnueli, A. (1977): The temporal logic of programs. Proc. 18th Ann. Symp. Foundations of Computer Science, Providence, RI. New York: IEEE, pp. 46–57

Pnueli, A. (1979): The temporal semantics of concurrent programs. In: Kahn, G. (ed.): Semantics of concurrent computation. Proc. Int. Symp., Evian. Lect. Notes Comput. Sci. *70*. Berlin-Heidelberg-New York-Tokyo: Springer, pp. 1–20
Pnueli, A. (1981): The temporal semantics of concurrent programs. Theor. Comput. Sci. *13*, 45–60
Prior, A.N. (1957): Time and modality. Oxford: Oxford University Press
Prior, A.N. (1967): Past, present and future. Oxford: Oxford University Press
Rescher, N., Urquhart, A. (1971): Temporal logic. Vienna-New York: Springer
Schlingloff, H. (1983): Beweistheoretische Untersuchungen zur temporalen Logik. Techn. Univ. Munich, Inst. for Informatics, Diplomarbeit
Segerberg, C. (1967): On the logic of tomorrow. Theoria *33*, 45–52
Shoenfield, J.R. (1967): Mathematical logic. Reading, Mass: Addison-Wesley
Wolper, P. (1983): Temporal logic can be more expressive. Inf. Control *56*, 72–99
v. Wright, G.H. (1965): And next. Acta Philos. Fenn. *18*, 293–304
v. Wright, G.H. (1966): And then. Commentat. Phys.-math. XXXII 7
v. Wright, G.H. (1968): Always. Theoria *34*, 208–221

Further reading

We give only a few papers to indicate the many fields in which temporal logic is applied.

Barringer, H., Kuiper, R., Pnueli, A. (1984): Now you may compose temporal logic specifications. Proc. 16th Ann. ACM Symp. Theory of Computing, Washington, D.C., pp. 51–63
Bernstein, A., Harter, K. (1981): Proving real-time properties of programs with temporal logic. Proc. 8th Symp. Operating Systems Principles, Pacific Grove, CA, pp. 1–11
v. Bochmann, G. (1982): Hardware specifications with temporal logic: An example. IEEE Trans. Comput. C-*31*, 223–231
Clarke, E.M., Emerson, E.A. (1981): Design and synthesis of synchronization skeletons using branching time temporal logic. In: Kozen, D. (ed.): Logics of programs, Workshop, Yorktown Heights. Lect. Notes Comput. Sci. *131*. Berlin-Heidelberg-New York-Tokyo: Springer, pp. 52–71
Emerson, E.A., Halpern, J.Y. (1983): "Sometimes" and "Not never" revisited: On branching versus linear time (preliminary report). Proc. 10th Ann. ACM Symp. Principles of Programming Languages, Austin, Texas, pp. 127–140
Gergely, T., Úry, L. (1980): Program behaviour specification through explicit time consideration. Proc. IFIP Congr. 1980, Tokyo. Amsterdam: North-Holland, pp. 107–111
Halpern, J., Manna, Z., Moszkowski, B. (1983): A hardware semantics based on temporal intervals. In: Diaz, J. (ed.): Automata, languages and programming, 10th Colloq., Barcelona (ICALP83). Lect. Notes Comput. Sci. *154*. Berlin-Heidelberg-New York-Tokyo: Springer, pp. 278–291
Lehmann, D., Shelah, S. (1983): Reasoning with time and chance. In: Diaz, J. (ed.): Automata, languages and programming, 10th Colloq., Barcelona (ICALP83). Lect. Notes Comput. Sci. *154*. Berlin-Heidelberg-New York-Tokyo: Springer, pp. 445–457
Lichtenstein, O., Pnueli, A., Zuck, L. (1985): The glory of the past. In: Parikh, R. (ed.): Logics of Programs, Proceedings, Brooklyn. Lect. Notes Comput. Sci. *193*. Berlin-Heidelberg-New York-Tokyo: Springer, pp. 196–218
Manna, Z., Wolper, P. (1984): Synthesis of communicating processes from temporal logic specifications. ACM Trans. Program. Lang. Syst. *6*, 68–93
Ramamritham, K., Keller, R.M. (1983): Specification of synchronizing processes. IEEE Trans. Software Eng. SE-*9*, 722–733

Reif, J., Sistla, A.P. (1983): A multiprocess network logic with temporal and spatial modalities. In: Diaz, J. (ed.): Automata, languages and programming, 10th Colloq., Barcelona (ICALP 83). Lect. Notes Comput. Sci. *154*. Berlin-Heidelberg-New York-Tokyo: Springer, pp. 629–639

Schwartz, R.L., Melliar-Smith, P.M. (1982): From state machines to temporal logic: Specification methods for protocol standards. IEEE Trans. Commun. Com-*30*, 2486–2496

Schwartz, R.L., Melliar-Smith, P.M., Vogt, F.H. (1983): Interval logic: A higher-level temporal logic for protocol specification. In: Rudin, H., West, C.H. (eds.): Protocol Specification, Testing and Verification III. Amsterdam: North-Holland, pp. 3–18

Subject Index